Origin on Trial

Origin on Trial

Big Evolution or Biblical Creation?

Christopher H. A. Ting

RESOURCE *Publications* • Eugene, Oregon

ORIGIN ON TRIAL
Big Evolution or Biblical Creation?

Resource Publications
An Imprint of Wipf and Stock Publishers
199 W. 8th Ave., Suite 3
Eugene, OR 97401

www.wipfandstock.com

PAPERBACK ISBN: 978-1-6667-4824-6
HARDCOVER ISBN: 978-1-6667-4825-3
EBOOK ISBN: 978-1-6667-4826-0

01/26/23

To Reiko, Eri, Ai, Joshua, my sisters, and their families.

Contents

Preface

MANY OF US FEAR science because science is difficult to understand. We leave the hard stuff to scientists. So we are always overwhelmed when world-class scientists say amazing things. "Nothing caused the Big Bang. Nothing."[1] Stephen Hawking continued, "The universe is the ultimate free lunch."[2] Carl Sagan said, "We are made of star-stuff."[3] And Richard Dawkins declares, "Evolution is a fact."[4] Famous scientists are so smart that we assume they can't be wrong.

Nonetheless, not all fields of science are created equal. There are two broad categories. One deals with how Nature behaves and operates *now*. The other is about specific events and processes in the *past*. Respectively, let's call them operational science and historical science.[5]

Our universe is either eternal or it had a beginning. Either way, mainstream big evolutionists give a very old age to our universe. So big cosmic evolution is about the past history of deep time.

We don't see big changes in biology occur every day. A monkey still reproduces another monkey without fail. But big bio evolutionists claim that an *ancient* common ancestor must have evolved into the monkey we see today and us. So big bio evolution is also about the past history of deep time.

We can conclude that Big Bang and big bio evolution are two areas of historical science. Together, they aim to solve the grand mystery of the origins of the universe, life, and biodiversity.

But do you know that there are dark secrets in these fields of historical science? For example, we are told that dinosaurs went extinct sixty five million years ago. It is really a long time. But among the fossils of dinosaurs, stretchable soft tissues continue to be found. Some of them even have dead

1. Hawking, *Answers*, chapter 1, para. 24.
2. Hawking, *Answers*, chapter 1, para. 21.
3. Sagan, *Cosmic*, 190.
4. Dawkins, "Theory," last para.
5. Batten, "It's," para. 14.

blood cells. Could these degradable biomaterials survive thousands of years, let alone sixty five million years?

Yet another example is diamond. It is said that natural diamonds took a billion years to form. But after accounting for the background noise, scientists still detect excess carbon-14 atoms in them. Similar to the shelf life of lettuce in the supermarket, carbon-14 has a short half-life. It should have all decayed away in six hundred thousand years. But we still find the radioactive carbon-14 in diamonds.

Another dark secret is Earth. Our planet is said to have originated four and a half billion years ago, according to the measurable rate of radioactive decay.[6] We also know that our Moon is moving away from us at a measurable rate. If we were to wind back the clock, as big evolutionists must always do, the Moon would be so close to Earth to kiss it, about one and a half billion years ago.[7] In other words, it's impossible for the third rock from the Sun to be older than this rendezvous date in the past.

Many historical scientists handwave away these unwelcome problems. Some have proposed a few ways to solve the enigma. For example, it is suggested that the speed of lunar recession must have been lower *in the past*.

But they can't go back in time to measure the *past* rate of lunar recession. Neither can they measure the *past* rate of radioactive decay of each isotope. They can only *assume*. Historical science has an intractable limitation. The initial conditions of the *real* origin are unknown. Big evolutionists have to push them into their models by hand, so as to fit the data collected today.

These empirical conundrums and limitation of big evolution prompt us to be wary when we come across the claims of historical scientists. The fame of a scientist may make the discovery sound credible. But any genuine discovery is repeatably testable by other scientists at present. This feature of being repeatably testable is a crucial attribute of operational science.

In contrast, historical events are unique and they are not even testable. The models of historical science may provide a nice fit with observations. And they do. There is no dispute about that. But this agreement with data doesn't automatically imply that the *modeled* events actually happened in the pathway of history. Many things in the distant past are unknowable. Historical scientists must speculate and believe.

• • • • • • •

6. The radioactive decay rate is not really a constant. See Jenkins et al., "Evidence," 42–46.

7. For a recent technical paper on lunar recession, see Green et al., "Explicit," 46–53.

For more than a thousand years, educated people of the Western world believed in Ptolemy's geocentric theory. It was the norm of their worldview. The theory fits some data pretty well. Yet, it has quite a few problems with the same set of data. In Ptolemy's model, the motion of every planet is circular and its speed doesn't change. But data show that the planetary orbits are elliptical and their speeds do vary. Some supporters of Ptolemy's theory tried to fix these problems. They introduced new parameters and made the model more complicated. However, their attempts failed. Ptolemy's theory is found wanting.

Today, the Big Bang model is in the same situation. We hear cosmologists speak of anomalies, tensions, and crises. Big Bang is not as rock solid as it is presented in textbooks and regurgitated by enthusiasts. The fact that Big Bang is sinking is a dark secret of modern cosmology.

I find it necessary to inform you about the serious problems of big evolution. A balanced view is overdue. Given the *crises* in historical science, it doesn't make sense for Christians to place historical science above the book of Genesis in the Bible. The word "Genesis" means origin. Genesis is the foundation of gospel. Christians need to realize that big evolution's origin is a rampage on Genesis.

• • • • • •

Christians are often criticized for failing to provide any *scientific* evidence that the Bible is divinely inspired. Even if Big Bang and big bio evolution are demonstrated to be flawed like Ptolemy's theory, it doesn't necessarily follow that biblical creation is true. This is a legitimate criticism and the argument is logical.

Therefore, in chapter 2, applying the method of data science, I set up a hypothesis test. I then use two formulas to test a pair of twin sentences in the Bible. These two verses proclaim the genesis of everything, and "in the beginning" God was there.

I have repeatedly tested the discovery of a triplet of three constants reported by Jenkins and his collaborators. With the same precision of five significant digits, they are pi, the base of natural logarithms, and the fine structure constant of quantum physics. Amazingly, these three numbers of Nature can be *derived* from Genesis 1:1 and John 1:1.[8]

8. If you are unfamiliar with the Bible, I recommend the online rendition of New American Standard Bible 1995, courtesy of the Lockman Foundation. The homepage is https://nasb.literalword.com/. You can search for, say Genesis 1:1, by typing "Genesis 1:1" in the interactive search box.

The reason this triplet is a scientific discovery is that these three constants are critically important to math and science, especially physics. For the triplet signal to be a fluke is much more improbable than the already improbable criterion physicists set to weed out weak signals.

Who would have expected to find these three constants embedded in the twin verses? Their writers probably didn't know these three numbers. With Occam's razor, most likely it was the divine Mind pushing the pen. Since this triplet signal spans the Old and New Testaments, it is reasonable to attribute it to the divine inspiration of God of the Bible—the Creator.

This verifiable result is a strong evidence that at least two verses in the Bible are divinely inspired. Their literal meanings indicate that Jesus is God and He is the Creator of heavens and earth. The three constants may well be the signature Jesus signed on His masterpiece.

Since Jesus is God the Creator, He speaks the truth and only the truth, for God cannot lie. Jesus affirmed the Old Testament. His words in the New Testament are obviously divinely inspired. Therefore, we have evidence to believe that the Bible in Hebrew and Greek is the work of the Holy Spirit.

Scientists, theorists, theologians, philosophers, scholars, popes, church fathers, pastors, and any other mortal people, have no greater authority than Jesus has. He is the final arbiter on matters in the Bible that overlap with the speculations of big evolutionists.

Did Jesus say He created a population of humans who could become sick and die before sin entered the world? Big evolutionists' model of ape-like "humans" struggling to survive are incompatible with the perfect couple God made in His own image. God gave Adam and Eve the Garden of Eden to enjoy, not the Hunger Games to suffer and die.

The divine wisdom of Jesus is far beyond the best human mind. The Holy Bible is the ultimate authority of the living God.

• • • • • •

I hope that you will examine the scientific arguments and the latest appraisals by historical scientists themselves. Science becomes not that difficult when we are as curious as a child. We just need a little bit of resolve to understand. We just need to keep asking questions.

To avoid committing the strawman fallacy, I have quoted verbatim from many writings by journalists, scientists, and atheists. The epithet of each chapter serves as its theme. Also, please do not skip reading these at times lengthy quotes, for I am avoiding the possibility of quoting them and the Bible out of context.

Each chapter is written to be as self-contained as possible. So you will find a few repetitions here and there. If you are only interested in proving me wrong, please jump straight to chapter 2. Christians who aren't sure whether the Bible is *really* the Word of God may find chapter 2 helpful in clearing their doubts.

Obviously, many important topics are omitted. Since many books have already covered big bio evolution, my focus is on big cosmic evolution.[9]

A key message is to urge everyone to discern carefully by asking questions. Is it empirical or theoretical? Is the science operational or historical? Is it observation or is it an interpretation of observation? Is it a fact or a claim? Is the claim repeatably testable? Is the claim experimentally reproducible?

I strive to help you see the half-truths from truths in science. Even so, as a former atheist, I know that ultimately it boils down to your heart. Do you want to continue to be god yourself? Or do you want to be at peace with the Creator and trust Him? He will give you the peace you need to have rest in your life.

9. For a comprehensive and updated articles on different areas of historical science, in no particular order, I refer you to NW Creation Network, Creation-Evolution Headlines, Answers in Genesis, Creation Ministries International, Apologetics Press, Institute for Creation Research, Origins Resource Association, and many others listed on the webpage https://creation.com/creationist-organizations-in-the-usa. I learn a lot from these resources.

Acknowledgments

I AM GRATEFUL TO Willie Chow, Joo Chye Young, and especially Jen Kai Chen for reading the drafts and their comments. I thank Reiko for her prayers and support. I also want to acknowledge all the people of the publisher who made this book possible in print. Big thanks to you! I am humbled by the grace of God for the undeserved like me to be useful in some way.

1. Introduction

> Science is incompetent to reason upon the creation of matter itself out of nothing. We have reached the utmost limit of our thinking faculties when we have admitted that because matter cannot be eternal and self-existent it must have been created.[1]
>
> —James Clerk Maxwell

1.1 Two Conflicting Worldviews

I WAS AN ATHEIST. I had an atheistic worldview. But not anymore.

I know some atheists see Christians as anti-science and incapable of thinking straight. These words are insulting. But they don't bother me because I know some atheists, rightly or wrongly, are angry at Christians, even God of the Bible.

Nevertheless, Jesus, God of the Bible, taught everyone the virtue of not returning evil with evil but good. Using hyperbole, Jesus gave us the wisdom of life:

> But I say to you who hear, love your enemies, do good to those who hate you, bless those who curse you, pray for those who mistreat you. Whoever hits you on the cheek, offer him the other also; and whoever takes away your coat, do not withhold your shirt from him either. Give to everyone who asks of you, and whoever takes away what is yours, do not demand it back. Treat others the same way you want them to treat you.[2]

Following this golden rule, I hope to engage in discussions with atheists, non-Christian, and Christians who don't share the same worldview. I will

1. Maxwell, "Molecules," 376.
2. Luke 6:27–31.

do my best to be respectful and intellectually honest. I will also avoid making logically fallacious arguments.

First and foremost, to set the record straight, most atheists don't really hate, curse, mistreat, and hit Christians on the cheek. But some atheists have an almost pious commitment to the worldview of big evolution. Some of these atheists become angry when Christian apologists, as politely as possible, raise doubts that strike the very core of big evolution.

The adjective "big" refers to the ultimate evolution from the Big Bang to you and me. Many scientists believe that the laws of Nature *only* can bring about big evolution, because Nature is all that is. Depending on the context, big evolution may refer to big bio evolution, or big cosmic evolution, or both.

Next, to prevent misunderstandings, like most people, Christians are generally not flat-earthers. Also, Christians are never anti-science. Never. On the contrary, the origin of *modern* science can be traced to Europeans who believed in God of the Bible.[3]

Most importantly, Christians never dispute the small changes observable in microorganisms. These living things have the ability to adapt to the changing environments. Be it natural or man-made, changes at the micro level are undeniable facts. Everyone knows that some bacterial strains can become resistant to antibiotics. We are also mindful of different varieties of orchids that look very different from each other. By trial and error, breeders of cats and dogs know how to produce a new breed.

Despite their adaptations, bacteria are still germs. After 75,000 generations,[4] Escherichia Coli bacteria are still E. Coli. These bacteria still make infected persons suffer severe stomach cramps and diarrhea. Despite the changes, orchids are still orchidaceous flowers. Cats are still felines, and dogs are still canines. This kind of small changes is generally known as microevolution. Most of the experiments in the lab motivated by evolutionism are dealing with microevolutions.

A more accurate description of these small changes is perhaps "adaptive mutation." This is because in most cases, changes are brought about by mutations. Sometimes, a genetic swing occurs and we observe a loss of biological functions.[5] Microevolution is not always progressive.

Adaptive mutation has no conflict with the biblical account. Sadly, the perfect creation became fallen as a result of Adam's sin.[6] In our imperfect

3. Wallace, *Person*, chapter 9.
4. Callaway, "Legendary," 634–635.
5. Pagel et al., "Loss-of-function," i389.
6. Genesis 3:17–19.

fallen world, we have many natural and man-made disasters. The Bible tells us the origin of these calamities. By contrast, big evolution offers no explanation. In this worldview, disasters are needed for big changes and extinction.

With a caveat, the notion of adaptive mutation agrees with Forbes and Krimmel's definition of microevolution. They define microevolution narrowly as "changes below the level of species."[7] The caveat is speciation in a controlled environment. Since such speciation with *small* samples in lab experiments is scientist-guided, it is unnatural. The artificially generated "new specie" (defined as, for example, not willing to mate) in these experiments should be considered as adaptive mutation also.

This caveat is also justified by the fact that "specie" is defined in many ways. John Wilkins, a historian of bioscience, listed over two dozen definitions of "specie."[8] Biologists have many species of "specie."

But intuitively, we can readily recognize a fruit fly from a housefly as different species. To name a new specie for fruit flies which refuse to mate seems a bit too academic than practical.[9]

Now, a big issue surfaces when scientists *extrapolate* adaptative mutation to macroevolution. It is *the* big evolution that requires new genetic information. Since adaptive mutation is a scientific fact, evolutionists claim that macroevolution must also be a scientific fact too. For example, Coyne, a leading evolutionist, writes,

> . . . there is nothing in either the theory or data from evolution implying that evolutionary change could be limited: as far as we can see, macroevolution is simply microevolution extended over a long period of time.[10]

Coyne's definition of macroevolution makes big evolution sound as if it is so ready to happen. Yet, Coyne wastes no time to invoke the assumption of "over a long period of time" for macroevolution to occur. If we indeed can observe big bio evolution today, this assumption will be utterly unneeded. But we don't observe any big evolution. A very long period of time is always needed and assumed in big evolution. It is the trump card—Darwin's god of the gap.

Note also that the theory of big bio evolution has zero power to predict the exact amount of time required for an unknown number of unknown adaptive mutations to produce a big evolution. According to many

7. Forbes and Krimmel, "Evolution," para. 7–9.

8. Wilkins, "Defining," sect. 11.

9. For a reference on species, see Jeanson, *Replacing*, 1–336.

10. Coyne, *Why*, 256.

evolutionists, "evolution does not progress toward an ultimate or proximate goal. Evolution is not 'going somewhere.' It just describes changes in inherited traits over time."[11] When starting points and end points are uncertain, it is obvious that nobody can ascertain the time, the number, and the type of adaptive mutations needed for a big evolution to happen.

The word "extension" used by Coyne is extrapolation, meaning that it's unobservable today. Evolutionists must assume that, given enough deep time in the past, complex living things could somehow emerge through some natural mechanisms.

Now, macroevolution isn't about the usual science of cause to effects. Rather, it is to see the effects (fossils, genes, etc.) today and then hypothesize their unobservable cause in the past, such as an unknown common ancestor. But no one has ever seen the common ancestor of monkey.

On the other hand, the Bible clearly declares that God, the Father, the Son, and the Holy Spirit created all the living things. Most crucially, Genesis and elsewhere in the Bible reveal that humans are *not* the results of big evolution. Sharply different from apes, we humans are living beings with the spiritual "breath of life."[12] And God loves us so much that He sent the Son of God to die in place of our sins and transgressions. Many atheists will respond, "Really?"

Whether atheists believe it or not, the Holy Bible reveals to us that atheists too are created in the image of God. The lives of atheists are also immeasurably more valuable than gorillas and chimps. But many atheists believe that, according to the *history* of big evolution, these primates shared a common ancestor with us once upon a time. Humans are therefore their distant cousins. And cosmologists tell us that we are all star dust. Let us admit it. We tend to believe whatever big evolutionists say with no question asked. Do Christians dare to respond, "Really?"

For Christians, the origin of biodiversity is supernatural in the beginning. But for atheistic and theistic evolutionists, some non-living things changed into plants. Some plants changed into animals. Some fish changed into creeping things on the ground. And some of these creeping things changed into birds.

But all these big changes are not observable because they all supposedly happened in the past—a very long time ago.

11. Forbes and Krimmel, "Evolution," para. 6.

12. Genesis 2:7 and 7:22. The word "breath" נְשָׁמָה (*neshamah*) appears twenty three times in the Old Testament. It is a figure of speech for human and the spirit of man. This divine *neshamah* is given exclusively to humans only. No animal has it. Therefore, made in the image of God, only humans worship God "in spirit and truth." (John 4:23).

We have one world but two very different versions of origin. Suppose we are given the same data "godisnowhere" from observation.[13] We may interpret it as "God is nowhere" followed by "to be found." Or, we may interpret it as "God is now here" followed by "with His people." Which is correct?

Concerning the origin, atheistic worldview of big evolution and biblical worldview of creation can't be more divided.

1.2 What is and isn't Science?

The word "science" is perhaps one of the most misunderstood and misused terms in our modern culture. Because science has produced so many wonderful innovations, it has been elevated to the status of a modern demigod.

Scholars are also not immune to the temptation of using this demigod to up the image of their fields of study. Some speak of hermeneutics as an art or *science* of interpretation. But this claim is debatable. For example, we can't resurrect Plato and ask him to clarify what he actually meant in some of his writings. Only Plato the writer himself can give us the model answer. Even so, Plato's answer is not science.

What then is science?

Science is a body of objective facts about Nature. These facts can be scientifically tested and re-tested by anyone, at almost any time, and almost everywhere. A definitive hallmark of science is repeatable testability.[14]

In particular, under the same set of conditions, everyone who knows how to perform the required operations, whether theoretical, experimental, or computational, should be able to replicate similar if not the same results independently, repeatably, and at present time.

The criterion of repeatable testability is extremely important.[15] As an illustration, let's consider an electric current flowing in a conductor. We know that even the best metal conductor will resist the current and generate heat. That's why our smart phones become warm after using them for heavy-duty work. This is an empirical *fact* that everyone can verify. That is, electrical resistance is verifiable, by almost anyone, everywhere, and every time.

Next, suppose a scientist makes an extraordinary claim. Under conditions ABC, an electric wire of material XYZ can conduct electricity without any electrical resistance at all. This phenomenon is called superconductivity.

13. Credit is due to Moses Gummadi for this "observation."

14. Some branches of psychology and economics deal with repeatably observable human perceptions and behaviors. They are rightly called the social science.

15. For a thorough discussion, see Stadler, "Scientific," chapter 2.

Given our knowledge of non-zero electrical resistance, nobody needs to apologize for being at first skeptical of this "super" claim.

Other competent scientists can test this "super" claim by performing experiments with the same conditions and material. If the same result is obtained over and over again, then superconductivity is said to be repeatably testable, and the claim is verified to be true.

Indeed, over and over again, superconductivity has been tested to be positive. Engineers have even constructed magnetically levitated trains with superconductivity. Therefore, we have a strong case to elevate superconductivity to the status of a scientific fact. In other words, we can include the discovery of superconductivity into the body of knowledge called science. Superconductivity is science.

On the other hand, if a claim of discovery is not reproducible, then it remains a hypothesis, a speculation, anything but a scientific fact.

A nice example is cold fusion—a cold version of the hot nuclear fusion of hydrogen inside every star such as the Sun. On March 23, 1989, two electrochemists made an extraordinary claim at a press conference.[16] It was before their accepted paper was published.[17] They announced that nuclear fusion could occur in their modest lab. If their discovery were true, cold fusion would provide the cleanest possible energy to mankind. They would become the candidates most likely to win the Nobel prize.

Many people tried to replicate or reproduce their results, but none succeeded. Recently, a group of scientists embarked on a major program to revisit the "cold case" of cold fusion. Their research efforts were also "yet to yield any evidence of such an effect."[18] Today, we don't consider cold fusion as science because it isn't reproducible. That's why we don't see cold fusion mentioned in any physics or chemistry textbook, except when it is introduced as a negative example of science.

Let's consider for a minute to grant that the two scientists did really observe cold fusion in their lab. But if all other scientists, including themselves, can't replicate the result, what is the value of that discovery? Nevertheless, their claim is at least testable. It turns out that their claim has been tested to be negative.

The orthodox experimental science has its focus on either establishing or disproving, objectively, a hypothesis or claim that something is the cause of the effect it produces. In the case of superconductivity, material XYZ under conditions ABC is the cause. The perfect flow of electric

16. Huizenga, *Cold*, appendix III.

17. Fleischmann and Pons, "Electrochemically," 301–308.

18. Berlinguette et al., "Revisiting," 45.

current is the resulting effect. In the case of cold fusion, the cause doesn't produce the expected effect.

Next, computational science is about designing and writing software for simulation. An example is the aerodynamics of wind resistance of tall buildings and long span bridges. The inputs to a computer simulation are data collected through scientific observations. The virtual-reality outputs of a simulation, however, aren't as one-hundred-percent reliable as those of the real-reality experiments.

What about data science? The "data" part of data science may be easy for us to wrap our mind around it. We know that observational science is essentially data collection. Astronomy is a good example. Astronomers design and build equipment and system to observe the sky for discovering new celestial objects and phenomena. They *operate* the telescope system to capture and record electromagnetic and gravitational signals. Likewise, in molecular biology, DNA sequencing and genome mapping are also observational science.

But where is the "science" part in data science? As much as material science is science, is data science a science too? According to IBM, data science is defined as

> a multidisciplinary approach to extracting actionable insights from the large and ever-increasing volumes of data collected and created by today's organizations. Data science encompasses preparing data for analysis and processing, performing advanced data analysis, and presenting the results to reveal patterns . . .[19]

The "science" part of data science, according to this definition, is arguably the patterns that are "revealed" by "advanced data analysis." If the results of data analysis reveal a pattern, in principle, comparable results that reveal the same pattern can be independently replicated. Other data scientists, using exactly the same data and algorithms, will produce patterns that are highly similar to it.

The algorithm or model can be implemented in different programming languages. For the same data, however, the results should be similar, and the same patterns should be revealed, regardless of operating system, computer hardware, etc. Otherwise, if the claimed pattern can't be reproduced, then it's most likely a false signal—a statistical illusion. Perhaps the spurious pattern is generated by some bugs in the computer codes.

We can never overemphasize the need to ensure that no statistical illusion occurs. Imagine what would happen when a decision or conclusion

19. IBM Cloud Education, "Data."

were made on the basis of such patterns that don't really exist. These false positives would give rise to misinformation and misrepresentation, which invariably lead to a loss of credibility. For some applications, severe financial losses might even occur.

With regard to applications in science, Blei and Smyth discuss data science from the perspective of scientific research. They describe it as "the child of statistics and computer science."[20] They identify three areas where classical paradigms need to be replaced by data science. These three areas are sequencing technology, digital sky surveys, and digitization of documents. We should also include the big data generated by the financial markets around the world.[21] Data science is well positioned to help scientists working in these domains. They can now take full advantage of their data.

In fact, data science has changed the way astronomers discover new objects in space. Zhang and Zhao review big data astronomy and machine learning tools, and how they are valuable in helping astronomers to process massive amount of images.[22] Remarkably, Kunimoto et al. report their discovery of seventeen new planets when they were combing through the big data from NASA's Kepler mission.[23]

1.3 Four Types of Confusion

From the contrasting examples of superconductivity and cold fusion, we can appreciate the importance of repeatable testability to distinguish science from non-science.

Now, it may be surprising to many people, even scientists. If we use the yardstick of repeatable testability, there are four types of confusion about what science really is.

First, we tend to confuse the use of scientific equipment and apparatus with the field of study itself. Observations and data collections done scientifically don't necessarily imply that the observed object is science. For example, using X-ray to scan a clump of charred parchment leaves in a lab[24] doesn't necessarily mean that the biblical contents deciphered from the scan are science.

Second, we tend to befuddle the availability of observations with the interpretation of observations. What most people—scientists

20. Blei and Smyth, "Science," 8689.
21. Ting, *Algorithmic*, 4.
22. Zhang and Zhao, "Astronomy," 1–9.
23. Kunimoto et al., "Searching," 1–20.
24. Wade, "Scanning," para. 1–3.

included—don't seem to realize is that observations need to be *interpreted* with a model. We also tend to confuse an interpretation of observation as a scientific fact. Most of the time it is, but sometimes it isn't. The second type of confusion is quite subtle.

A good example to elucidate this type of confusion is spectral lines. Each element in the periodic table has its own set of wavelengths when the atom absorbs a packet of energy called photon, which is light. This absorption produces a series of black lines in the spectral graph. They look like barcodes. They also function like barcodes for astronomers to scan the contents of elements in a star or a galaxy.

The main composition of a star is hydrogen. We measure the spectral lines of hydrogen for the Sun, and likewise for a distant galaxy. We expect the two spectral graphs of hydrogen to be identical. But in fact, they aren't the same. Compared to the wavelengths of our Sun, the ones from a galaxy far away are either slightly longer or shorter. The shift to a longer wavelength is called the redshift. The shift to a shorter wavelength is known as the blueshift. The spectral shift is then quantified as the ratio of the shifted wavelength over the unshifted one measured from the Sun.

Most of the times, we observe redshifts. This part is observation. The other part is interpretation. In the Big Bang model, redshift is *interpreted* as the expansion of space itself by an *unknown* dark energy. This interpretation dominates the scientific community to the point of being an unquestionable doctrine. But redshifts can be explained by some other physical mechanisms that don't require dark energy.

For example, a recent paper shows that redshifts can be explained by the quantum physics of electromagnetic attenuation as light traverses the inter-galactic vacuum of quantum fluctuations.[25] Another example interprets redshifts with the mechanism of variable velocity of light.[26] Yet another example is a finding that correlated fluctuations within a galaxy can produce redshifts.[27] Which interpretation is correct?

The third type of confusion has to do with the practice of replacing "logical and objective" by the word "scientific" as a way to suggest trustworthiness. Performing an investigation study can be "scientific" but the object of study itself may not be science. A task-force committee carries out an investigation "scientifically" but the object under investigation, say, the misconduct of certain individual scientists, is not science. Similarly, forensic

25. Musha, "Alternative," 26–31.

26. Petit, "Interpretation," 1527–1532.

27. Wolf, "Non-Cosmological," 363–365.

experts gather pieces of circumstantial evidence "scientifically" from the crime scene but surely crime is not science.

Some scholars think that history is also a science. But by definition, historical events don't repeat as we wish. We can use scientific equipment to measure accurately, say, the elevation of a historical site. Nonetheless, the event in history that provides historical significance to the site can't be observed. Take the historical crucifixion of Jesus as an example. Does it make sense to let Jesus die and resurrect a few times for us to observe the mechanism of resurrection?

The last type of confusion is very common and very important to highlight. We often fail to realize that not all fields of science are created equal. It is extremely crucial to differentiate those that deal with the present from those that look into the past. The former is non-historical operational science. It is the science of how Nature *operates* on the daily basis. The same laws of Nature that govern how it *operates* apply in the first century as well as in the twenty first century. They have no regard for history.

On the other hand, non-operational historical science is about the origin of the universe, the origin of the solar system, and the origin of life. It is historical because these origins happened in the past, at least a few thousand years ago. If there is such thing as the physics of the origin, it doesn't work today. In contrast to planets and comets, the origin of the universe won't happen again for us to observe. Origins are unique events that happened only *once* in the past, exactly like history. We can't observe past origins.

In the purely materialistic worldview of big evolution, both Jesus and the universe are materials. Obviously, both the resurrection and the Big Bang can't be observed today. We can't study the mechanics of resurrection. By the same token, it's also impossible to let the universe be born again for scientists to confirm their speculations. History doesn't repeat itself. Big Bang and common ancestors won't happen again for us to observe.

It's very important to differentiate history from science.

Christians admit that believing in gospel writers' *eyewitness* accounts of Jesus is faith through the work of the Holy Spirit.[28] By contrast, many people tend to think that believing in what cosmologists tell us about the origin of the universe *without* eyewitnesses is not faith but science. This

28. Many books on the historical reliability of resurrection have been published. See, for example, Strobel, *Case*. Another accessible book is Wallace, *Cold-Case*. These two authors *were* atheists. They wanted to prove Christianity wrong. But their investigations led them to change their verdict. They are Christians now. Believing in Jesus's resurrection is never a blind faith. This event is historically *falsifiable*. All the opponents in the first century had to do is to show the corpse of Jesus, which was under their control. Nobody could find the dead body of Jesus after three days of His death.

is a common confusion from not distinguishing operational science from historical science.

Observation data of the universe are the effects we have today. They serve as evidence for the *true* cause that had already happened. Making claims on a *possible* historical cause that is consistent with the effects is historical science. But a possible cause may not be the true cause. We have many suspects. But only one of them is the "culprit."

Of course the true cause is unique. The origin of the universe happened only once. Therefore, when we use the criterion of repeatable testability, cosmology of early universe isn't really a blue-chip science. It's a version of history.

Trying to construct the history of our natural world since the origin of the solar system is another example of historical science. Although historical scientists are "scientific" in their various speculations, they can't know for sure how did the solar system *really* begin to exist. Nobody has seen the formation of Earth with the eyes.

1.4 Big Bang is not Observable

Some cosmologists use their models to suggest that the universe could come from "nothing" by chance.[29] But nobody has ever observed a universe like ours born from "nothing." Cosmologists weren't around to see its birth with their eyes.

A *theoretical* possibility written on a few pieces of paper doesn't mean that the *physical* origin did really happen with that possibility. An example similar to this situation is the Kaluza-Klein theory. It is a theory that unifies Einstein's gravity with Maxwell's electromagnetism in five-dimensional spacetime. But is the Kaluza-Klein theory a description of reality?

Since ancient times, philosophers are always proposing their models to fathom the origin of the cosmos. From the online description[30] of *Ancient Greek Cosmogony*,[31] I quote,

> The atomists argued that our world was one among many worlds, and came about by chance. Plato argued that it is unique, and the product of design. Empedocles and the Stoics, in quite different ways, argued that there was an unending cycle whereby the world is generated, destroyed and generated again.

29. Krauss, *Universe*, 1–224.
30. Gregory, "Description," para. 2.
31. Gregory, *Ancient*, 1–314.

> Aristotle on the other hand argued that there was no such thing as cosmogony,[32] and the world has always existed.

Intriguingly, atomists' cosmogony is strikingly similar to the modern fashion of multiverse from "literally nothing."[33] Our universe is just "one accident among many accidents."[34] Plato's argument for uniqueness and design is akin to the modern school of intelligent design. The philosophical framework of Empedocles and the Stoics is almost the same as the paradigm of cyclic models proposed by Penrose. Finally, Aristotle's cosmology is similar to the static model preferred by Einstein.

In other words, purely naturalistic theories of the origin, fanciful as they are, had already been considered thousands of years ago by philosophers. There is no shortage of theory for the origin of the universe. Theory is cheap but the reality is rich. In essence, nothing is new under the sun.

Suppose a famous theoretical physicist claimed that the expansion of the universe was due to a super-giant black hole outside the universe. Would you believe?

Using scientific methods, nobody can prove the black-hole idea right, and nobody can prove it wrong either. Why? No human can travel outside the universe to observe. When a claim isn't scientifically testable, it is a philosophical speculation. By the same token, nobody can travel back in time to observe the events at the beginning. Nobody has ever repeatably observed the beginning from "nothing." The origin is not observable.

You may protest and say, "Observations of astronomy can be best data-fitted by an inflationary Big Bang model." You are right.

But from the history of science, we learn that Ptolemy's geocentric model was widely accepted for many years. It can describe and predict the positions of the planets. The predictions fit the observation data known as the Alfonsine Tables very well.[35] Yet, the geocentric model is deficient. Who knows, perhaps the widely accepted Big Bang model is like Ptolemy's model?

Even so, Ptolemy's model and the Big Bang model are fundamentally different. The former deals with planets and the Sun, which are directly and repeatably observable. The latter deals with big evolution. Nobody has ever

32. Strictly speaking, a theory on the origin of the universe is cosmogony. But in today's context, cosmology is defined as a branch of astronomy that deals with the origin.

33. Vilenkin, "Creation," 26. The word "nothing" is a misleading misnomer. The model of Vilenkin assumes the gravitational field, matter field, and nonzero energy density. Surely these are something.

34. Gregory, *Ancient*, 2.

35. Britannica, "Alfonsine," line 2–3, and Gingerich, *Book*, chapter 4.

observed the Big Bang, nor the birth of a galaxy. As cosmologists always assume, they happened billions of years ago.

In the court of law, observations for the purpose of historical science aren't eyewitness accounts. Why? Observations of astronomy are *not* carried out at the time of the origin. Also, a model with various assumptions and many parameters is always needed to interpret the data.

Likewise, repeatable experiments and observations are impossible for the origin of the solar system. The Sun is unique. We can't replicate its origin in the lab. When a speculative model of natural history can't be repeatably tested, it is a story.

An unfussy way to understand this point is to ask a simple question politely for Big Bang cosmologists to answer. Do you have a photograph of the instant when the Big Bang occurred? Do you have a video footage of the process of our planet forming itself?

The cosmologists might respond that these are silly questions. They will tell you instead that they have observed this and that today. However, when you persist, some cosmologists may admit that they don't have any observation of any of the origin. The origin had already happened. They can't observe at the time of origin.

The nice inflationary Big Bang illustrations such as the one in Figure 1.1 that we often see are not photographs. All the beautiful big evolution illustrations and videos are generated by computer graphics. There is nothing wrong with using digital *arts* for illustrations. But it must be emphasized that they aren't the direct observations of our universe's origin. They aren't the processes of cosmic evolution.

By contrast, in day-to-day operational science, there is only one objective way to interpret the measurements or data collected for, say, an electric current in a conductor and the voltage applied across it. Ohm provided a simple relation between these two physical variables, which has become a scientific fact known as Ohm's law. It states that the voltage is proportional to the electric current by way of the electrical resistance. Everyone performing the same experiment of Ohm will surely come to the same conclusion. At room temperature, Ohm's law is always true. Ohm's law is not history. Ohm's law is science.

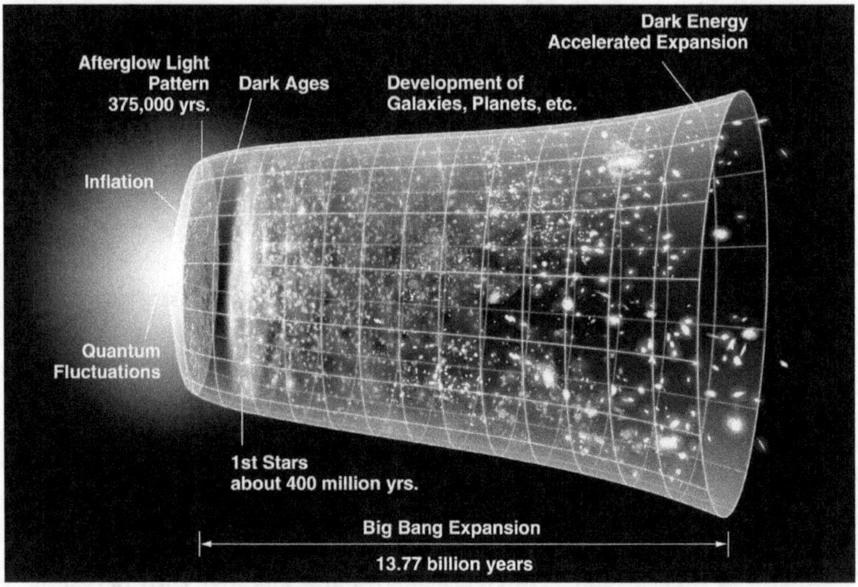

Figure 1.1. A digital artist's computer graphics of inflationary Big Bang, which is the consensus model of today's cosmology. Credit: NASA/ WMAP Science Team and Cherkash.

1.5 Science and Miracles

The four types of confusion lead many people to mistake history as science. With "science" being a strongly positive and desirable word to associate with, we tend to find the possibility of big evolution more plausible than biblical creation. So, many lean toward the natural *history* written by scientists seeking to explain various origins with the laws of Nature *only*. The lack of exercising critical thinking makes people overlook the complete lack of direct observation of big evolution in action.

Many Christians accept big bio evolution.[36] According to Davis, a core tenet of theistic evolution is the following.

> The Bible is NOT a reliable source of scientific knowledge about the origin of the earth and the universe, including living things—because it was never intended to teach us about science.[37]

36. Martin, "Compatibility," 1–20.
37. Davis, "Theistic," para. 10.

Theistic evolutionists confuse the history of origin as science. They seem to view the illustration of Big Bang in Figure 1.1 as a hard fact. Like most people, they mix up historical science with operational science.

Of course, the Bible never intended to teach us about superconductivity, redshift, adaptive mutation, and so on. These are operational science.

But the Bible does intend to teach us about history. Starting from the *origin* of time, the Bible reveals to us the origin of Earth, the origin of water on Earth, the origin of day and night, the origin of season, and the origin of living things. All things came into being through Him, and apart from Him nothing came into being that has come into being.[38]

1.5.1 Miracles and the Supernatural Realm

Perhaps the easiest way for Christians to justify this mentality of "historical science first" is to argue that we shouldn't read the first eleven chapters of Genesis literally.

Some theologians argue that the events and people in these opening chapters are allegories. According to the Merriam-Webster online dictionary, allegory is "the expression of truths or generalizations about human existence by means of symbolic fictional figures and their actions. It encompasses such forms as fable and parable." In the allegorical interpretation, God's six-day creation is not to be taken literally as an ordered series of historical events.

Some may suggest that we should leave it to scientists working on big evolution to tell us with their observations and descriptions, for scientists are always right. However, this suggestion is the result of suffering from the four types of confusion discussed earlier. Sometimes, scientists are wrong. The philosophy of "historical science first" is a form of scientism.

To interpret the opening chapters of Genesis other than a historical account goes against the basic principles of hermeneutics. To begin with, as many as forty five verbs in Genesis chapter 1 have the conjunctive ו (*waw*) prefixed to each of them. Grammatically, the heavy use of these time-sequential verbs strongly suggests that the first two chapters of Genesis, like the rest of the forty eight chapters, provide a historical narrative. These two chapters are not poems.

Classifying the opening chapters of Genesis as an allegory is extremely damaging to the Christian faith. Allegorical interpretation denies that God is the *hands-on* Creator. This interpretation also contradicts various passages in the Bible, which consistently attest to God's hands-on creation.

38. John 1:3.

It is necessary at this juncture to raise the awareness of the issues caused by the mentality of "historical science first" rather than "Bible first." What are the implications when the six-day creation is interpreted allegorically? Why can't we think that big evolution can be wrong?

1.5.2 Biblical Worldview

First, we all know that a human dead for two to three days can't become alive again. If self-described Christians say they don't believe Jesus's physical resurrection because of this scientific fact, they will become heretics. As it is written in a book of the New Testament,

> Now if Christ is preached, that He has been raised from the dead, how do some among you say that there is no resurrection of the dead? But if there is no resurrection of the dead, not even Christ has been raised; and if Christ has not been raised, then our preaching is vain, your faith also is vain. Moreover we are even found to be false witnesses of God, because we testified against God that He raised Christ, whom He did not raise, if in fact the dead are not raised. For if the dead are not raised, not even Christ has been raised; and if Christ has not been raised, your faith is worthless; you are still in your sins. Then those also who have fallen asleep in Christ have perished. If we have hoped in Christ in this life only, we are of all men most to be pitied.[39]

Apostle Paul who wrote this book in the sixty-six-book Bible strongly argued for Jesus's resurrection as the foundation of Christian faith. If there was no physical resurrection, Christians would be "testifying against God," and "of all men most to be pitied." It's game over.

Next is Mary's virgin birth of Jesus. What if a Christian says the virgin birth is unscientific?

We all know that making a baby is impossible without intercourse. Joseph the husband of Mary understood this scientific fact too. Before an angel of the LORD revealed to him, Joseph initially thought that Mary had an extra-marital affair. "Being a righteous man and not wanting to disgrace her, [Joseph] planned to send her away secretly."[40] But Joseph had a biblical worldview. He believed what an angel of God said in a dream, and he followed the instructions the angel gave him.

39. 1 Corinthians 15:12–18.

40. Matthew 1:19.

Can Christianity stand if Jesus was born in the same way as us, with the proclivity to sin? If Jesus were merely a man, He would have to atone for His own sin, "for all have sinned and fall short of the glory of God."[41] His sacrifice on the cross would be futile, being blemished by sin and thus unacceptable in the sight of the holy God. Christians' beliefs would become trash, "vain," and "worthless;" Christians would be "still in sins."[42]

If Jesus were just an ordinary man, Jesus wouldn't be the only begotten Son. It means that Jesus wouldn't be from God. The doctrine of Jesus as a mere ethics professor is "anti-Christ."[43] Although Christians in the first century were reminded to defend Christian faith "with gentleness and reverence,"[44] there are times when they called a spade a spade.

The two central doctrines (resurrection and virgin birth) of Christianity require a biblical worldview. As illustrated in Figure 1.2, in the biblical worldview, the natural world is in coupling union with the supernatural realm. At His own will and power, God can temporarily suspend the laws of Nature at a pinpoint location and time. From our point of view, it is a miracle.

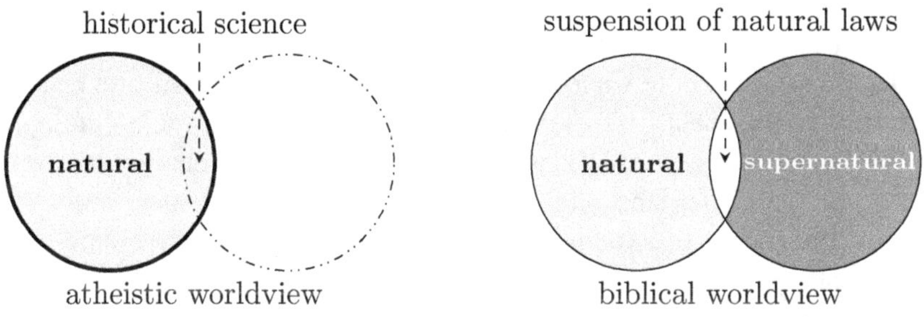

Figure 1.2. Comparison of two worldviews.

If we accept the possibility of supernatural suspension by God's sovereign will and power, then it isn't difficult to believe in the miracles of virgin birth and resurrection. It then becomes very natural to understand why Jesus could perform so many miracles, including turning water into wine, walking on water, and calming the stormy sea. Jesus just spoke, and Nature obeyed. Why? He is the Creator. He knows everything about Nature. God is the ultimate scientist.

41. Romans 3:23.
42. 1 Corinthians 15:17.
43. 1 John 4:1–6.
44. 1 Peter 3:15.

Biblical creation is a miracle. It is a piece of cake for God to create. God knows more about the Law of Nature than Einstein. Why? He wrote it. He is the Law Giver.

1.5.3 Why must God be the Hands-on Creator?

Having laid the groundwork for the biblical worldview, we are now ready to examine the far reaching implications of interpreting the beginning portion of Genesis allegorically.

First, we ask, "How can God suspend the laws of Nature?" Well, it must be that He was the only One who designed and created the universe along with the laws of Nature in the first place. He is the only One who knows how to do it.

To understand this crucial point, imagine for a minute that you have designed and created many high-tech pharmacy stores. You have the knowledge of medicine and the high-tech know-how to manage many stores remotely from your office.

To promote your brand, you hire a few celebrities to buy things from one of your stores. You obviously will suspend your day-to-day business in that chosen store only, while other stores run normally. You will also make special arrangements so that the celebrities are the only customers during the time of televised visit. You will be there to answer their questions. After the visit, the chosen store returns to its normal operation.

The point is that you must be the founder and owner of the stores to suspend one of them. Otherwise, you have neither the operational expertise nor the legal power to disrupt the business operation as you wish. Therefore, Jesus must be the executive Creator of both Nature and the laws that govern it. God has power over Nature. And Nature obeys His command.

Perhaps that's why in the apostles' creed, every Christian is reminded to first "believe in God, the Father almighty, Creator of heaven and earth." Being the Creator, God is also the owner. Throughout the Bible, God clearly made it known to all that He is the Creator and owner from the beginning.

If God weren't the "hands-on" Creator, He wouldn't even know how to suspend the laws of Nature at a pinpoint spacetime. God wouldn't be God if He couldn't perform any miracle. It would mean that Jesus was just an ordinary man born by natural conception, and His resurrection would have to be a make-up story. Then Jesus wouldn't be the Messiah either. Christianity would be an absolutely false religion.

Therefore, biblical creation isn't a secondary doctrine. It's as fundamental as the physical resurrection of Christ.

1.5.4 How did Jesus Interpret Genesis?

It is no coincidence that the Bible begins the biblical history with God's supernatural creation, with Adam as the real grown-up—not baby.[45] According to the gospel of Luke, Adam, the first man in history, was "the son of God."[46] Sadly, through Adam, "sin entered into the world, and death through sin, and so death spread to all men, because all sinned."[47] If Adam were a fictional figure, there would be no sin in the world, and Christianity would be false. I would remain as an atheist.

Genesis was written for the original audience of ancient Hebrews wondering in the wilderness. They would most likely interpret literally what Moses had written about what God the LORD had done for them, including the creation.[48] They had literally experienced the awesome presence and power of God, much more than we in the twenty first century can ever understand and imagine.

Genesis is the foundation of Israel. Genesis is also the foundation of Christianity. Jesus Himself affirmed it. His apostles and disciples also affirmed Genesis as history.[49]

Can Christians disbelieve Jesus's affirmation of Genesis? As written in the gospel according to John, Jesus told the unbelieving Jews,[50]

> For if you believed Moses, you would believe Me, for he wrote about Me. But if you do not believe his writings, how will you believe My words?[51]

45. This interpretation is consistent with Adam being able to understand God's command (Genesis 2:16–17), and to give names to "all the cattle, and to the birds of the sky, and to every beast of the field." (Genesis 2:20). Adam could even compose a short poem (Genesis 2:20) on day six of creation when he first saw Eve.

46. Luke 3:38.

47. Romans 5:12.

48. Literal interpretation doesn't mean that we must interpret that God has a mouth and a tongue like us to speak. We need to recognize idioms and literary devices in our expository reading when we interpret the texts plainly and straightforwardly, i.e., literally. This is a common practice when we read any text.

49. Lisle, *Genesis*, chapter 2.

50. John 5:18. To prevent misunderstanding of anti-Semitism, the word "Jews" in this context doesn't refer to the Jews in Israel and all over the world. Rather, it refers to a specific group of Jews seeking to kill Jesus. Jews are still God's chosen people. Jesus was born to a Jewish family. It is natural and biblical for Christians to love His chosen people and pray for them. Shalom.

51. John 5:46.

What did Moses write in the book of Genesis? Moses wrote about God the LORD. Here, in no uncertain terms, Jesus was claiming that He was God the LORD, who created the heavens and the earth.

Therefore, interpreting the beginning portion of Genesis as an allegory is unbiblical. Treating Adam as a non-historical character is a disbelief in Jesus.

1.5.5 Biblical History versus Natural History

Some theorists go so far to conclude that "nothing" can create the universe all by itself. This assertion implies that God wasn't needed. So God doesn't even have to exist. This assertion is a direct rejection of the biblical account. It is utterly incompatible with Christianity. Logically either God exists, or God doesn't exist.

We need to recognize that, in contrast to operational science, the origin of the universe isn't a day-to-day operation. God's six-day creation is purely supernatural. Genesis 1 provides the history of supernatural creation. "And God spoke," "and it was so." No natural processes known today apply to this description of God's creative acts.

Also, importantly, most atheists will agree that *natural* history is incompatible with Genesis not only in the nature of creation but also in the details of creation. Genesis says that "God created the heavens and the earth," and the earth's surface was in the beginning covered with water. The apostle Peter interpreted the creation and Noah's flood in Genesis literally. He wrote, ". . . by the word of God the heavens existed long ago, and the earth was formed out of water and by water . . ."[52]

But according to natural history, the Earth was naturally formed by gravity and electrostatics. "The newly forming Earth was incredibly hot," "characterized by a partially molten surface . . ."[53] Can we interpret water as incredibly hot magma?

In addition, God made stars on day four after He created Earth on day one and the dry land on day three. But in the history of big evolution, stars were formed billions of years before Earth. Also, birds were created on day five before the land-based dinosaurs on day six. But big evolutionists say, "No, some land-based dinosaurs evolved into birds." Again, on day five, God created the great sea monsters, and then all the land-based animals on day six. But big evolutionists say, "No, some land-based animals evolved into whales."

52. 2 Peter 3:5–6.

53. Johnson et al., *Introduction*, chapter 8.

Atheistic and biblical versions of natural history can't be more incompatible. Trying to harmonize the sequence of creation in Genesis with big evolution requires rewriting the Scripture.

When we try to accommodate one part of the Bible with certain parts of big evolution, the supposedly superior interpretations always contradict major parts of the Bible. Intellectual dissonance occurs when we resort to sophistry.

1.6 What is Next?

The above arguments are somewhat brief. You may not be convinced. But in the subsequent chapters, from the standpoint of science, I hope you will journey with me to understand the severe flaws of big evolution. You will get to see that Big Bang is in big trouble.

Few people know about the serious problems in historical science. Many scientists seldom mention the dark secrets of big evolution. We can understand why they don't like to talk about it.

Well, if you want to continue your stance of "historical science first," I respect your decision. Me too, I had previously believed in what famous scientists of natural history say about the origin for many years.

But new observations of astronomy show that the flaws in historical science are really severe. I can't continue to believe what historical scientists say without doubts anymore. A growing number of scientists are looking for new models, even a complete overhaul. Why should Christians stick to big evolution that is replaceable?

Note that my views on the current models of origins have little to do with my born-again conversion to Christianity from atheism. I am merely asking questions with critical thinking. What I doubt only are ideas, models of natural history, and the *interpretations* of scientific observations. I trust that you can judge whether my objections are factual and rational.

Also, it is never my intention to doubt even an iota the scientific integrity of almost all historical scientists. I admire their drive and creativity. We do science because we love it. We spend hours after hours thinking—at times agonizing. We do many arduous experiments and go on field trips to harsh and dangerous places like Darwin did to obtain samples. These are hard work. Scientists are among the most hardworking people, all for the love of science.

2. Does the Bible have the Signatures of God?

The secret things belong to the LORD our God, but the things revealed belong to us and to our sons forever, that we may observe all the words of this law.[1]

It is the glory of God to conceal a matter,
But the glory of kings is to search out a matter.[2]

Numero pondere et mensura Deus omnia condidit.
God created everything by number, weight and measure.[3]

—ISAAC NEWTON

2.1 Two Conflicting Positions and a Logical Fallacy

MANY BIBLE CRITICS REGARD the Holy Bible as a book of myths. It has nothing worthy of thoughtful consideration in terms of historicity and reality. To critics, the Bible is just another book; and Genesis is similar to Hesiod's *Theogony*. To put it bluntly, Bible is just a bunch of fairy tales. It isn't holy, and Jesus is not God.

Take for example Joseph Sommer, an activist of humanism. He published an article on the website of American Humanist Association. In the opening, he wrote,

> Humanists reject the claim that the Bible is the word of God. They are convinced the book was written solely by humans in an ignorant, superstitious, and cruel age. They believe that because

1. Deuteronomy 29:29.
2. Proverbs 25:2.
3. Robertson and O'Connor, "Quotations."

> the writers of the Bible lived in an unenlightened era, the book contains many errors and harmful teachings.[4]

This view of the Bible is similar to what a New Testament scholar Bart Ehrman is championing. Sommer and Ehrman are hardly the lone voices in the wilderness. The Pew Research Center reports that, in 2014, thirty three percent of American adults think that the Bible, written by men, is not the word of God.[5] Also, fifty five percent of Jews regard the Old Testament in the same way.[6]

Negative view of Christianity is nothing new. Ancient Roman historians didn't have any respect for biblical beliefs. For example, Tacitus described Christianity as a "mischievous superstition."[7] It implies that the Bible is erroneous and void of truths.

On the other hand, Christians take it as a matter of doctrine and faith that the Bible contains the words of God. Being superintended by the Holy Spirit, biblical writers authored their books for their intended audiences. The Bible isn't one book but a library of sixty six books of many authors. They are written in the original languages of Hebrew with some Aramaic, and Greek. Each book contains the matters God intended to speak to His people since eternity.

The atheistic and biblical positions on the Bible can't be more contrasting.

The doctrine of inerrant and infallible Bible in the original languages is founded on the Bible itself. It teaches that the whole Bible contains the words of God. His words are truths because "God cannot lie."[8]

This doctrine of inerrancy may seem to be a circular reasoning. But it isn't. Apostle Paul, a well-trained scholar, wrote in a letter to Timothy to affirm that all *Scripture*, written many centuries earlier by many authors, "is inspired by God."[9] Suppose you find a book on a bookshelf affirming the contents of thirty nine ancient books on another bookshelf. Logically, it's wrong to describe this affirmation as circular reasoning just because they are in the same library. The Bible isn't one book but a library of sixty six books written by many different authors over many centuries.

Before proceeding further, we need to point out a logical fallacy in Sommer's writings quoted above. Living in an "unenlightened era" doesn't

4. Sommer, "Reasons," line 1–4.
5. Cooperman et al., "U.S. Public," 58. See also Newport, "Fewer."
6. Cooperman et al., "U.S. Public," 58.
7. Tacitus, *Annals*, book 15.
8. Titus 1:2.
9. 2 Timothy 3:16–17.

necessarily imply that the Bible has many errors. Historically, the Romans are known for their "unenlightened" cruelty, ruthlessness, and beliefs in countless gods. Does it necessarily mean that *The Annals* written by Tacitus contains many errors?

The adjective "cruel" applies not only to ancients but to modern people as well. The methods to violently end the lives of fellow human beings cruelly have just become technologically advanced (e.g., the gas chambers of Holocaust, atomic bombs, abortions, mass shootings, etc.). But the outcomes are always the same: unnatural, violent, and painful deaths. Murderers are still cold-blooded and cruel.

Also, modern people are still superstitious. Some examples include friggatriskaidekaphobia (phobia for Friday the thirteenth), good luck for picking up a penny, bad luck for opening an umbrella indoors, knocking twice on wood to reverse bad luck, horoscope, and the list goes on. In fact, Vyse, a leading psychologist and superstition researcher, concludes that

> Despite the growing dominance of scientific reasoning, superstition, pseudoscience, and magical thinking did not go away... they may never go away.[10]

Three hundred years later, will people in the future say that what we wrote today contains "many errors and harmful teachings" because of ignorance, cruelty, and superstition in our society?

Therefore, we have reasons to argue that associating the Bible and Christianity *by default* with superstition is misinformation. This alleged connection to superstition is nothing new because with superiority complex, ancient Romans two thousand years ago thought so too. But whether a book contains errors or not must be based on evidence—both internal and external. It has nothing to do with whether we live "in an ignorant, superstitious, and cruel age."

Filtering away this misinformation, Sommer's opening three sentences can be objectively distilled into a position statement as follows. "Humanists think that the Bible is not the word of God, and the Bible has many errors." On the other hand, the Chicago Statement on Biblical Inerrancy claims the exact opposite. "The Bible is the word of God, and the Bible has no error." Simply put, it's about the Bible being "another book written by humans without any divine element at all" versus "the book written with divine inspiration."

10. Vyse, *Superstition*, chapter 4.

2.2 A Hypothesis Test and Five Sigma

Which of these two opposing positions is more reasonably believable? Rather than arguing over the definitions of error, God's words, and so on, we can answer this question by way of an objective test. Using statistics, we can capture Sommer's position as a null hypothesis:

H_0 : The signal of divine element in the Bible is null.

Statisticians use the word "null" to indicate zero value in the measurement. Translated into plain English, the null hypothesis is saying that there is no sign of divine element in the Bible.

This null hypothesis is patterned after the one put up by experimental physicists:

H_0 : The signal of Higgs particle in the experimental data is null.

What is the Higgs particle? It is an extremely short-lived particle that the standard model of particle physics predicts that it must exist. Its nickname is the God particle. The nickname became popular, thanks to the title of a book written by a science writer Dick Teresi and an experimental particle physicist.[11] He was Leon Lederman, an atheistic Jew and a Nobel laureate.

For years, experimental physicists have been conducting high-energy, high-precision experiments to collect a large amount of data in search of the Higgs particle. After several decades of research, experimental particle physicists finally found enough signals from the background data to reject the null hypothesis at the threshold of the so-called 5 sigma. A news release on CERN's official website reads:

> Both ATLAS and CMS gave the level of significance of the result as 5 sigma on the scale that particle physicists use to describe the certainty of a discovery. One sigma means the results could be random fluctuations in the data, 3 sigma counts as an observation and a 5-sigma result is a discovery.[12]

As shown in Figure 2.1, sigma (σ) of statistics is a standard measure of distance from the average value. Specifically, sigma is the distance from the null, i.e., zero value—no signal. The two shaded areas under the bell-shape curve in Figure 2.1 add up to approximately five percent of the entire area. These two symmetric and equal-size areas are called the rejection regions.

11. Lederman and Teresi, *God Particle*, preface.

12. CERN, "Higgs," para. 3.

On the other hand, the vast majority area of about ninety five percent, which contains the null value, is called the non-rejection region.

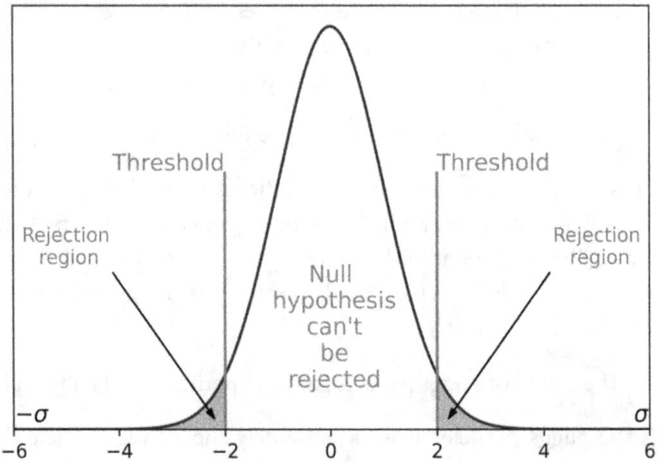

Figure 2.1. The area under the entire bell-shape curve is one hundred percent. The threshold is set at 2 sigma on the right and -2 sigma on the left. The area for each of the two symmetric rejection regions is close to 2.5 percent.

When a statistic for the signal computed from the data falls on the non-rejection region in Figure 2.1, it means that the null hypothesis can't be rejected. It is probably true. In the interval between -2 sigma to 2 sigma, we are about ninety five percent confident that the null hypothesis is true.

An illustration is the smoke detector. When the level of smoke in a room is below the threshold, we may say, "There's no signal." The test for a fire in a hotel room is negative.

Obvious as it may seem, the alternative hypothesis of the null hypothesis is that the signal is *not* null. When the computed statistic falls on one of the two rejection regions, the null hypothesis must be rejected. We consider instead the alternative hypothesis as a more probable scenario. The smoke detector gives a signal for smoke from a fire, as its value is not null. The test for a fire in a hotel room is positive.

When we shift the two threshold bars in Figure 2.1 to plus minus 5 sigma respectively, the hurdle to reject the null hypothesis becomes *exponentially* much more difficult. The non-rejection area becomes much larger, and the two symmetric rejection regions under the graph shrink so much that they become invisible in Figure 2.1. Now, we only have a 1 in 3.5 million chance for the noise in the background to look like a signal. Stringent high thresholds minimize false fire alarms.

This statistical approach is what the experimental particle physicists were taking as they went about searching for the elusive Higgs particle. As alluded to earlier, in July 2012, two independent teams of experimentalists at CERN announced that they had found a signal in their experimental data as it cleared the 5-sigma threshold. Since then, repeatedly, the Higgs particle continues to be "discovered." Higgs particle has become science.

Experimental physicists are also after the super-partner particles predicted by the theories of supersymmetry. So far, no signal has been detected. The null hypothesis can't be rejected, which means that no supersymmetric particle has been observed. Not only that, certain features of supersymmetry are inconsistent with experimental data. The paradigm of supersymmetry is still a speculation. Nonetheless, this speculation is at least experimentally testable, or falsifiable, as some people like to call it.

Another example of a negative test result is the existence of a magnetic monopole. Analogous to positive and negative charges, a magnetic monopole is an isolated magnet of exclusively either a north pole or a south pole. The grand unified theories predict that monopoles must be out there. But so far, none is found; the null hypothesis of no signal can't be rejected.

Using the same method, we take the words in the Bible as data and examine whether there is a signal to reject the null hypothesis. If the biblical data do not provide any signal at all to reject the null hypothesis, then Sommer has a compelling case to claim that, objectively, the Bible is just another book.

On the other hand, if the null hypothesis has only a tiny probability to be true, then it is to be rejected. Rejection of the null hypothesis is equivalent to the antithesis that the alternative hypothesis is more likely to be true, meaning that a signal is in the data. Still arguing that the Bible is just another book becomes unreasonable. It is like arguing that the Higgs particle hasn't been found.

We have contrasting cases of rejection and non-rejection of three respective null hypotheses for the Higgs particle vis-à-vis super-partners and magnetic monopoles. The bottom line is that a low threshold of 2 sigma makes it not difficult to reject Sommer's null hypothesis. It is one out of twenty two.

Therefore, to be rigorous and robust, let us adopt the high standard of experimental physicists: at least 5 sigma. The higher is this value of the nonlinear distance from zero, the exponentially harder it will be to reject the null hypothesis. Our goal is to make it ridiculously difficult to dismiss Sommer's assertion that the Bible is *entirely* uninspired.

2.3 Two Similar Number Systems

Before we can conduct a statistical test, we need to introduce two highly similar number systems for encoding the Hebrew and Greek words in the Bible, which are our data. You can easily devise the same scheme for English.

Today, we use the ten Arabic numerals 0 to 9. But it wasn't so in the past. Each people group has its own way to represent numbers. In the Jewish culture, the twenty two Hebrew letters were used for numbers. Even today, scriptural verse numbers in the Hebrew Old Testament, or the Jewish Tanakh, are based on the Hebrew letters.

2.3.1 Ancient Hebrew Number System

Table 2.1 presents the Hebrew number system in correspondence with the Arabic numerals. As Hebrew is a language that reads from right to left, the sequence of Hebrew letters in Table 2.1 follows the same convention.

Like English A to Z, starting from the first letter *aleph* א to the last letter *tav* ת, the ordinal sequence runs from 1 to 22. Notice that for the eleventh, thirteenth, fourteenth, seventeenth, and the eighteenth positions of the sequence in Table 2.1, each has a pair of letters. The left letter of the pair is used instead when a particular Hebrew word ends with that letter.

Table 2.1. Hebrew number system. The shaded rows contain the characteristic values of the letters.

11	10	9	8	7	6	5	4	3	2	1
כ ך	י	ט	ח	ז	ו	ה	ד	ג	ב	א
20	10	9	8	7	6	5	4	3	2	1

22	21	20	19	18	17	16	15	14	13	12
ת	ש	ר	ק	צ ץ	פ ף	ע	ס	נ ן	מ ם	ל
400	300	200	100	90	80	70	60	50	40	30

Let us now focus on the shaded rows of Table 2.1.[13] They give a numerical value for each letter. For the first ten letters, their ordinal position numbers and their numerical values are the same. However, for the next nine letters—from the eleventh to the nineteenth—the numerical values

13. Ifrah, *Universal*, 215. See also Powell, *Introducing*, box 30.2.

progress in tens. Finally, for the last three Hebrew letters, their numerical values jump by a hundred. It is obvious that this numerical scheme from antiquity is not arbitrary. There is a pattern.

With this system, for example, the number 73 can be encoded by the sixteenth and the third letters, i.e., עג. In other words, we add the numerical values of these two letters to obtain the number 73. From right to left, the letter *ayin* ע corresponds to 70, while *gimmel* ג is 3. By addition, this combination of two letters represents 73.

If we apply this letter-number system to each word in the Scripture, the natural consequence is that every letter and every word in the Old Testament will have a numerical value. Let's call it the characteristic value. It is simply the sum of all the numbers for the constituent letters of a Hebrew word.

Take the word חכמה (*chokmah*) for example. It is mostly translated as "wisdom." Its characteristic value is 73. From right to left, we have 8 for *het* ח, 20 for *kaf* כ, 40 for *mem* מ, and 5 for *hey* ה. The sum of these four numbers is 73. Respectively, these four numbers are at the eighth, eleventh, thirteenth, and fifth positions of the ordinal sequence in Table 2.1. What we can also do, is to add up these ordinal numbers. In this example, when we add up 8, 11, 13, and 5, we obtain 37, which is called the ordinal value for חכמה (*chokmah*).

What is interesting about the Hebrew word חכמה (*chokmah*) is that 73 is the mirror image of 37. It turns out that 73 and 37 are prime numbers. A prime number is a number that is larger than 1 and it can't be wholly divided by any number other than itself.

Prime numbers are incredibly special in number theory. They are indispensable in the RSA algorithm, which is the encryption we use today. Prime numbers are also interesting in Bible study, both in terms of their values per se, as well as their positions in the list of prime numbers. For example, 73 is the twenty first prime number, and 37 is the twelfth prime number. Again, we see that in relation to the order of prime numbers, we have another pair of mirror images: 21 vis-à-vis 12. Such mirror symmetry, a less fanciful name for "chiastic structure," is quite prevalent in the Bible.

Now, from the following verses, it is clear that wisdom חכמה (*chokmah*) and creation are very much connected.

a. Jeremiah 10:12 and repeated exactly as Jeremiah 51:15

   It is He who made the earth by His power,
   Who established the world by His wisdom;
   And by His understanding He has stretched out the heavens.

b. Proverbs 3:19

   The LORD by wisdom founded the earth,

By understanding He established the heavens.

Since by wisdom the LORD founded the earth, wisdom must have already existed before creation. Indeed, Proverbs 8:22–31 speak of wisdom before the beginning of the world. This internal consistency within the collection of sixty six books makes it crystal clear that wisdom is inseparably connected to creation.

2.3.2 Ancient Greek Number System

The twenty seven books of New Testament were originally written in Greek. Is there anything similar to the Jewish numerical system? As you can see from Table 2.2, the Greeks too, since ancient time that in fact predates the Jewish numerical scheme, had a system called isopsephy for representing numbers.[14]

There were twenty seven Greek letters in antiquity. But the letters for 6, 90, and 900 had become obsolete well before the first century. These obsolete letters don't appear in all the twenty seven books of the New Testament. Notice also that sigma has two forms. The one on the right in Table 2.2 is used when a word ends with sigma.

Table 2.2. Greek isopsephy. In ancient Greek, there were twenty seven letters. Three letters, however, had become obsolete. They are digamma ϝ for 6, koppa ϙ for 90, and sampi ϡ for 900.

1	2	3	4	5	6	7	8	9	10	11	12	13	14
α	β	γ	δ	ε	ϝ	ζ	η	θ	ι	κ	λ	μ	ν
Α	Β	Γ	Δ	Ε	Ϝ	Ζ	Η	Θ	Ι	Κ	Λ	Μ	Ν
1	2	3	4	5		7	8	9	10	20	30	40	50

15	16	17	18	19	20	21	22	23	24	25	26	27
ξ	ο	π	ϙ	ρ	σ ς	τ	υ	φ	χ	ψ	ω	ϡ
Ξ	Ο	Π	Ϙ	Ρ	Σ	Τ	Υ	Φ	Χ	Ψ	Ω	Ϡ
60	70	80		100	200	300	400	500	600	700	800	

Like its Hebrew counterpart, the numerical scheme in Table 2.2 for Greek isn't arbitrary too.

Some may argue that we should remove the archaic letters and renumber from one to twenty four with the numerical values from one to

14. Ifrah, *Universal*, 220. See also Powell, *Introducing*, box 30.2.

six hundred. But Table 2.2 is the one used by ancient people, as can be seen from, for example, *The Lives of the Twelve Caesars* by Suetonius. He was a Roman historian, and a contemporary of New Testament writers.

Suetonius recorded an intriguing rumor suggesting that Nero murdered his mother: Νέρων ἰδίαν μητέρα ἀπέκτεινε (*Neron idian mehtera apekteine*). It is translated as "Nero slew his mother."

There are two letters of nu (N and ν) in Nero's Greek name Νέρων (*Neron*) in the nominative form. Together we get 100. Then add it to 5 for epsilon έ, 100 for rho ρ, and 800 for omega ω to obtain a total of 1,005.[15] This Greek characteristic value (CV) is equal to the total CV of ἰδίαν μητέρα ἀπέκτεινε (*idian mehtera apekteine*), which is translated woodenly as "his mother slew."

You can use Table 2.2 to calculate that the CV of ἰδίαν (*idian*) is 75, μητέρα (*mehtera*) is 454, and ἀπέκτεινε (*apekteine*) is 476. The total of these three characteristic values is 1,005. Thus we have "Nero = his mother slew." A pun was intended.

As you can see, this mapping of words into numbers has nothing to do with magic, mysticism, necromancy, witchcraft, or anything occult.

Even so, it isn't wise for well-meaning Christians to use this kind of numerology to "prove" that the Bible is divinely inspired. We all agree that neither Suetonius nor people who spread the rumor were inspired by the Holy Spirit. Yet they could produce the same CV of 1,005 for Nero and for the phrase "his mother slew." Also, it is difficult to compute the probability for two strings of words that share the same CV.

2.3.3 Calculate the Number

Before moving to the statistical test in the next section, it is necessary to address a concern of applying numbers on biblical texts. Is this practice unbiblical?

By His sovereign will, God chose Hebraic men of various backgrounds to write all the sixty six books. Some argue that Luke wasn't a Jew, but the arguments aren't strong enough to be certain. In any case, the sixty six books were written mainly *to* and *for* Hebrews. They were the originally intended audience. Therefore, reading the Bible with the Hebraic perspective is important for interpreting biblical passages correctly and accurately.

Jewish hermeneutics has four layers. For our purpose, the first two layers are sufficient.

15. Suetonius, *Lives*, book 6, sect. 39.

The first layer, called פשט (*peshat*), which means to strip off, is to extract the plain, straightforward, and literal meaning of a given text. We are to strip ourselves of all consciously known views that may skew our plain reading. Being literal doesn't mean that we ignore idioms, figures of speech, genre, syntax, and context.

The second layer, called רמז (*remez*), which means to wink, is the allusive inference of something not directly stated in the text. This category may involve cross-referencing texts from other parts of the Bible.

The intent is to widen and deepen the scope of the given text. The category of *remez* follows the principle of using certain parts of the Bible to interpret or illuminate a given biblical text.

In Jewish hermeneutics, the usage of numbers to draw out the unobvious contents that enrich the message of the given text is a legitimate practice. It is a tool that is complementary to cross-referencing.

The first layer of *peshat* has no dispute. That is how we read any book anyway. However, a little clarification of using numbers in *remez* is perhaps needed.

It is widely recognized that 7 is a symbol of divine perfection.[16] On the seventh day, God rested. The book of Revelation has many sevens. Other examples include Jesus said to Peter, "I do not say to you, up to seven times, but up to seventy seven times."[17]

We also find the repetition of three sets of 14 in Matthew's genealogy of Jesus. Jewish readers of Matthew would immediately see the significance of 14. It is the characteristic value of David's name דָּוִד. Matthew put much weight on Jesus being the "son of David" as a fulfillment of God's promise to David. Moreover, 14 is uniquely a product (2 × 7) of two prime numbers.

We also know from several passages in the Bible that 3 is symbolic of Trinity, God in three persons. And three disciples were eyewitnesses of the transfigured Jesus in conversation with Moses and Elijah. "For there are three that testify: the Spirit and the water and the blood; and the three are in agreement."[18]

Now, all three synoptic gospels record Satan tempting Jesus.[19] Satan quoted a passage from the Bible to lure Jesus to jump from a high place.[20] Does Satan's quoting of a biblical text make it repulsive for Christians to

16. Farbridge, *Studies*, chapter 5
17. Matthew 18:22.
18. 1 John 5: 7–8.
19. Matthew 4:1–11, Mark 1:12–13, and Luke 4:1–13.
20. Psalm 91:11–12.

quote biblical verses? Obviously, Satan's misuse of a biblical text doesn't mean that Christians are to shy away from quoting the Bible.

Likewise, there are several organizations in the world bearing the name of "church" such as the Unification Church. The misuse of "church" by non-Christians shouldn't discourage Christians from naming their congregation as so-and-so church.

The same can be said of Hebrew and Greek letter-number systems. Their misuses are facts. But these unsavory abuses shouldn't turn Christians away from their legitimate applications.

From the vantage point of data science and natural language processing, every word has to be given a number for computational analyses to be possible. The most important matter is to stick to the principle of consistency with the *peshat* meaning of the given text.

In fact, the apostle John even encouraged Christians to calculate the number.

> Here is wisdom. Let him who has understanding calculate the number of the beast . . .[21]

2.4 Statistical Test

Having discussed the ancient systems for encoding Hebrew and Greek words, we are now ready to perform the test. I emphasize again that this test is totally different from the kind that links Nero to the phrase "his mother slewed." A reason for this emphatic claim is that a probability can be unambiguously computed in our statistical analysis. Furthermore, the test doesn't involve finding words that fit the bill.

Now, in particle physics, a theoretical model is needed to let experimental physicists know what particles to search for in the data recorded by the detectors. Similarly, we need a theological model to tell us what to look out for in the data set, which consists of about 31,000 verses.[22]

A major theological tenet of the Bible is that God is the ultimate Creator. He started time, and then in the beginning, God created the universe and everything in it. By contrast, no scientist can tell us how time started. It is thus natural to zoom into the very first sentence of the Bible, "In the

21. Revelation 13:8.

22. See an online table at Blue Letter Bible, https://www.blueletterbible.org/study/misc/66books.cfm.

beginning God created the heavens and the earth."[23] Is this verse on creation divinely inspired?

The very first verse of the Bible is the proclamation that God is *the* Creator of the physical world. In Hebrew, Genesis 1:1 has 7 words and 28 letters. The appearance of 7 and a multiple of 7 raises the prospect that perhaps a signal is concealed in the first verse of the Bible. So, we compute the characteristic value (CV) of each word based on the numerical scheme in Table 2.1. The results are presented in Table 2.3.

Table 2.3. Genesis 1:1. In the beginning God created the heavens and the earth.

the earth	and	the heavens		God	created	In the beginning
הארץ	ואת	השמים	את	אלהים	ברא	בראשית
296	407	395	401	86	203	913

Hebrew goes from right to left. The numbers in the last shaded row of Table 2.3 are obtained in the same way that the CV of 73 is calculated for wisdom חכמה (*chokmah*).

For example, the second word from the right ברא (*bara*) for "created" has a CV of 203, being the sum of 2 (for *beth* ב), 200 (for *resh* ר), and 1 (for *aleph* א). Notice also that the object marker את (*at*) with the CV of 401 has no equivalent word in English and many other languages. The sixth word ואת consists of ו (*waw*) for "and" and the object marker את (*at*). Thus, its CV is 407.

2.4.1 The Transcendental Number Pi (π)

In math, pi (π) is the ratio of the circumference of a circle to its diameter. That is, when the length around the edge of a circle is divided by the diameter, pi is the number that ensues.

At four decimal places, equivalently, five significant figures, pi is well approximated by 3.1416. Modern computing technology allows us to calculate pi to fifty trillion decimal places. Even so, the number of decimal places of pi is infinite.

In many fields of math, pi is a particularly important constant. In physics, it appears in Einstein's general theory of relativity. At the scale of nanometer, which is one billionth of a meter, a particle's dual behavior of

23. Genesis 1:1.

wave becomes salient. Waves are intrinsically linked to trigonometry where pi plays a crucial role. So quantum physics also needs pi.

Einstein's theory of general relativity for the cosmic scale and quantum mechanics for the nano scale are the two foundations of physics. Both need pi. In other words, pi is related to creation "in the beginning" when God instituted the laws of science to govern His created world.

Therefore, a theological "prediction" is that we should find pi in the first verse of the Bible about "the beginning."

Let's now perform the data analysis. First, we compute the product of the 7 words' characteristic values in Table 2.3. The result is

$$\text{product of 7 words} = 296 \times 407 \times 395 \times 401 \times 86 \times 203 \times 913$$
$$= 304{,}153{,}525{,}784{,}175{,}760.$$

Next, we compute the product of the numerical values for the 28 letters of Genesis 1:1 from right to left as follows:

$$\begin{aligned}\text{product of 28 letters} = {} & 400 \times 10 \times 300 \times 1 \times 200 \times 2 \times \\ & 1 \times 200 \times 2 \times \\ & 40 \times 10 \times 5 \times 30 \times 1 \times \\ & 400 \times 1 \times \\ & 40 \times 10 \times 40 \times 300 \times 5 \times \\ & 400 \times 1 \times 6 \times \\ & 90 \times 200 \times 1 \times 5.\end{aligned}$$

The first line in this expression is the product of six letters that spell the first word "in the beginning." And the last line is the product of five letters for "the earth."

The result is 23,887,872,000,000,000,000,000,000,000,000,000.

We can now calculate the ratio between 28 times the product of 28 letters and 7 times the product of 7 words as follows:

$$(28 \times \text{product of 28 letters}) / (7 \times \text{product of 7 words})$$
$$= (28 \times 23{,}887{,}872{,}000{,}000{,}000{,}000{,}000{,}000{,}000{,}000{,}000) /$$
$$(7 \times 304{,}153{,}525{,}784{,}175{,}760)$$
$$= 3.1416 \times 10^{17}.$$

This ratio is the result of as if we have a huge circumference divided by a large diameter. The computed ratio is expressed in the scientific notation whereby we can easily find that there are five significant figures, i.e., 3.1416.

These digits match pi exactly at this level of approximation.[24] The error is only 0.0012 percent.

It is quite surprising that pi is embedded in this first verse of the Bible. God declares that in the beginning He created the heavens and the earth. As pi is present in the two foundations of physics (quantum mechanics and general relativity), the concealed pi is consistent with the plain textual meaning of Genesis 1:1—the creation of a *physical* world and the laws of the science that govern it.

The foundational laws were discovered in the twentieth century. They all require pi. Paul Dirac, a great physicist, remarked, "God used beautiful mathematics in creating the world."[25]

A slight detour may be interesting. Was Paul Dirac an atheist or a theist?" Which God was he referring to? Dirac was an atheist when he was a young man. But according to Abraham Pais, a physicist and a close friend of Dirac, he seemed to have become a theist.

In writing a memorial article on Dirac, Pais contacted Dirac's widow. She wrote a letter back. "Paul was no atheist. Many times did we kneel side by side in Chapel, praying. We all know, he was no hypocrite."[26] This testimony from the most intimate person of Dirac is consistent with Dirac's later writing. He wrote, "God is a mathematician of a very high order, and He used very advanced mathematics in constructing the universe."[27]

Now, the very first verse of the Bible is remarkable. Not only does it encode pi, Genesis 1:1 also displays many surprising structures of symmetry. In a peer-reviewed paper,[28] Jenkins examines Genesis 1:1. He finds many symmetries in the lexical structure and also in the characteristic values. Jenkins finds various geometrical symmetries in this verse.

Critics may argue that it's just a coincidence. But to be credible in the objection, critics need to provide a concrete example of finding pi in another natural language to which this opening verse is translated. Can critics obtain pi as accurate as at five significant figures in, for example, English?

Instead of verbal arguments, we should look at the probability. Let's compute the probability of this "coincidence." The first significant digit "3" of 3.1416 has a one in nine chance of being correct since it can't take the value of 0 by the definition of scientific notation. The second to the fifth

24. Jenkins, *Second*, 84. It is unclear who originally found pi in Genesis 1:1. Most likely it is Jenkins and his collaborators.

25. Pagels, *Cosmic*, 295.

26. Pais, "Paul Dirac," 29.

27. Dirac, "Evolution," third last para. It would be rare for an atheist to use the reverent "He" rather than the grammatically correct "he" in the middle of a sentence.

28. Jenkins, "Ultimate," 184–196.

significant figures each has a chance of one in ten. Since every digit is independently taken, the odds for getting 3.1416 by random chance is therefore one in ninety thousand. This probability is lower than that of tossing a fair coin sixteen times, when each time it's a head. Try this experiment and see whether you get a run of sixteen heads.

This probability of one in ninety thousand for 3.1416 corresponds to 4.39 sigma.[29] It is less than 5 sigma required by physicists. Nonetheless, it can be counted as an observation since it is larger than 3 sigma. But at this stage, Sommer is right to claim that the signal isn't strong enough to reject the null hypothesis.

2.4.2 The Transcendental Euler's Number (e)

Another verse that contains "in the beginning" before our world began is John 1:1. "In the beginning was the Word, and the Word was with God, and the Word was God." This verse is the theological twin of Genesis 1:1. Being the first verse of John, it is highly unlikely for scribal errors to occur. We are reading the very exact words penned by apostle John.

A pioneering Greek philosopher Xenophanes wrote, "In the beginning the gods did not at all reveal all things clearly to mortals, but by searching men in the course of time find them out better."[30] This philosophical utterance pales in comparison to John 1:1's "In the beginning."

Unlike the gospels written by Matthew, Mark, and Luke, the gospel according to John is unique. This gospel focuses on who Jesus really *is*. Theologically rich contents constitute a large portion of John's gospel. It reveals to us that the Word, Jesus, *was* already there in the beginning before the creation.

Apostle John tells us in chapter 1 that the Word (*logos* λόγος) is Jesus, because "the Word became flesh, and dwelt among us, and we saw His glory, glory as of the only begotten from the Father, full of grace and truth."[31] We also learn from this verse that John and his fellow apostles were eyewitnesses of Jesus.

In the beginning, the pre-incarnate Jesus as the Word (*logos* λόγος) was with God, and He created the world, since "all things came into being through Him, and apart from Him nothing came into being that has come

29. A Python program to find sigma from probability and vice versa is available upon request.

30. Xenophanes, "Fragments," 71.

31. John 1:14.

into being."[32] Moreover, we also read that "the world (*cosmos* κόσμος) was made through Him."[33]

Table 2.4 presents the seventeen Greek words of John 1:1 along with their characteristic values based on the Greek isopsephy in Table 2.2. We find the word *logos* (λόγος) appears three times in this verse. Its characteristic value is 373, being the sum of 30 for lambda λ, 70 for omicron ο, 3 for gamma γ, another 70 for omicron ο, and 200 for the word-ending sigma ς.

Next, notice the grammatically required small letter iota (ι) in the second word ἀρχῇ (*archei*) beneath the last letter eta (η). Therefore, this Greek word which means "beginning" has five letters.

Table 2.4.1. John 1:1. In the beginning was the Word, and the Word was with God, and the Word was God.

In	the beginning	was	the	word	and	the	word	was
Ἐν	ἀρχῇ	ἦν	ὁ	λόγος	καὶ	ὁ	λόγος	ἦν
55	719	58	70	373	31	70	373	58
with		God	and	God	was	the	word	
πρὸς	τὸν	θεόν	καὶ	θεὸς	ἦν	ὁ	λόγος	
450	420	134	31	284	58	70	373	

The product of the 17 characteristic values in Table 2.4 is as follows:

product of 17 words =

$$55 \times 719 \times 58 \times 70 \times 373 \times 31 \times 70 \times 373 \times 58$$
$$\times 450 \times 420 \times 134 \times 31 \times 284 \times 58 \times 70 \times 373.$$

The result is 949,302,241,399,839,184,297,857,869,760,000,000.

John 1:1 is composed with 52 letters. So we compute the product of 52 letters' numerical values from left to right as follows.

product of 52 letters

$$= 5 \times 50 \times 1 \times 100 \times 600 \times 8 \times 10 \times 8 \times 50$$
$$\times 70 \times 30 \times 70 \times 3 \times 70 \times 200 \times 20 \times 1 \times 10$$
$$\times 70 \times 30 \times 70 \times 3 \times 70 \times 200$$
$$\times 8 \times 50 \times 80 \times 100 \times 70 \times 200 \times 300 \times 70 \times 50$$
$$\times 9 \times 5 \times 70 \times 50 \times 20 \times 1 \times 10$$

32. John 1:3.
33. John 1:10.

$$\times 9 \times 5 \times 70 \times 200 \times 8 \times 50$$
$$\times 70 \times 30 \times 70 \times 3 \times 70 \times 200$$
$$= 8.43625145625911083008 \times 10^{75}.$$

This product is also written exactly.

Finally, using the same formula for obtaining pi, we compute the ratio between 52 times the product of 52 letters' characteristic values and 17 times the product of 17 words' characteristic values as follows:

$$(52 \times \text{product of 52 letters}) / (17 \times \text{product of 17 words})$$
$$= (52 \times 8.43625145625911083008 \times 10^{75}) /$$
$$(17 \times 949{,}302{,}241{,}399{,}839{,}184{,}297{,}857{,}869{,}760{,}000{,}000)$$
$$= 2.7183 \times 10^{40}.$$

Though tedious, these straightforward, transparent, and verifiable calculations yield a number 2.7183. Interestingly, this five-digit number corresponds to an approximation of another important constant of math—Euler's number, which is denoted by *e*.[34] Also known as the base of natural logarithms, the constant *e* appears practically everywhere.[35] For example, the ubiquitous bell-shape curve of statistics in Figure 2.1 is drawn by using *e* as the base for the exponential function.

Using the same division formula, the result for John 1:1 matches the Euler number 2.7183 exactly at five significant figures. The error is only 0.0011 percent.

Critics may dismiss this outcome as another coincidence. But it is difficult for critics to explain why John 1:1's "in the beginning" about the pre-incarnate Jesus being with God in creation would produce a constant of math. This constant is needed in describing many aspects of the physical world. It is consistent with the plain meaning of this beginning verse on God creating our physical world.

Together with the odds of getting pi, the probability for correctly obtaining two sets of five significant figures each is now down to 1 in 8.1 billion. This probability is significantly lower than the criterion of 1 in 3.5 million used by experimental physicists to announce the discovery of a particle. Our signal from two channels is 6.43 sigma from zero. It clears the 5-sigma threshold by a long shot. It can't be just another coincidence. It should be counted as a scientific discovery since π and *e* are both heavily used in natural science and social science also. I need *e* for my research in quantitative finance.

34. Jenkins, *Second*, 87. It is unclear who originally found *e* in John 1:1. Most likely it is Jenkins and his collaborators.

35. Reichert, "e," 982.

Therefore, we have reasons to remark that there is something very unusual in these two verses. Both reveal to us the origin of time and the origin of the universe. We can logically conclude that the revelation by God and the *logos* about biblical creation has to be taken seriously.

Now, Sommer can't claim that the Bible is entirely uninspired anymore since the null hypothesis is robustly rejected. At least two verses pertinent to "the beginning" were very specially written as if by the "hand of God."

To refute the validity of this discovery, critics are welcome to produce pi and *e* from all natural languages that say something meaningful and declarative. For example, "The universe is the ultimate free lunch."[36] Another instance is "Nothing in biology makes sense except in the light of evolution."[37]

2.4.3 The Fine Structure Constant (α)

Is there more? Yes. First, you can readily verify that the addition of seven characteristic values of Genesis 1:1 in Table 2.3 is 2,701. Likewise, for John 1:1, we compute the sum of seventeen characteristic values in Table 2.4 to obtain 3,627.

We concatenate these two sums by the order of their appearances in the Bible to make a new number 27,013,627. We then multiply this number with itself. So we obtain 729,736,043,695,129. Expressing this number in the scientific notation with five significant figures, we obtain 7.2974×10^{14}.

Intriguingly, these five significant figures match exactly those of a very important constant in quantum physics—the fine structure constant.[38] Denoted by α, the fine structure constant has been measured accurately. It tells us the strength of interaction between an electron and a photon—between matter and light.

It is remarkable that our computed value differs from the measured value by only 0.0001 percent.

The reciprocal $1/\alpha$ is rather close to 137. Every physicist knows the significance of 137. But nobody knows why it is dimensionless, and why it must be this value. The fine structure constant is a deep mystery of operational science. Below is what a great physicist Richard Feynman said about the fine structure constant, which he called the coupling constant.

36. Hawking, *History*, 147.

37. Dobzhansky, "Nothing," 125–129.

38. Jenkins, "Themes," sect. 5. It is unclear who originally found the fine structure constant from Genesis 1:1 and John 1:1. Most likely it is Jenkins and his collaborators.

> There is a most profound and beautiful question associated with the observed coupling constant, α—the amplitude for a real electron to emit or absorb a real photon . . . It has been a mystery ever since it was discovered . . .
>
> Immediately you would like to know where this number for a coupling comes from: is it related to pi, or perhaps to the base of natural logarithms? Nobody knows. It's one of the greatest damn mysteries of physics: a magic number that comes to us with no understanding by man. You might say the "hand of God" wrote that number, and "we don't know how He pushed His pencil." We know what kind of a dance to do experimentally to measure this number very accurately, but we don't know what kind of a dance to do on a computer to make this number come out—without putting it in secretly![39]

Our analysis shows that Feynman's conjecture of a possible relation of the fine structure constant to pi and *e* ("the base of natural logarithms") is spot on. Yes, indeed, the fine structure constant is obtainable from two unusual verses in the Bible that have pi and *e* embedded respectively in each of them.

Feynman was a man of physics, and a great teacher of physics. More than anyone, he didn't take the concepts of physics for granted. For example, in the famous *Feynman's Lectures*, he said, "It is important to realize that in physics today, we have no knowledge of what energy is . . . It is an abstract thing in that it does not tell us the mechanism or the reasons for the various formulas."[40]

Surely Feynman was an atheist. Surely Feynman was not joking when he wrote in a bestseller,

> I agree that science cannot disprove the existence of God. I absolutely agree. I also agree that a belief in science and religion is consistent.[41]

Feynman also thought that "Western civilization stands by two great heritages." He wrote, "One is the scientific spirit of adventure . . . The other great heritage is Christian ethics—the basis of action on love, the brotherhood of all men, the value of the individual, the humility of the spirit."[42]

39. Feynman, *QED*, 129.

40. Feynman et al., *Feynman Lectures*, last para. in sect. 4-1.

41. Feynman, *Meaning*, 36. Before critics claim that I quote Feynman out of context, I ask them to read the entire passage in this book of Feynman.

42. Feynman, *Meaning*, 47. It is a paradox that Feynman, an atheist, found Christian ethics pivotal. But again, people change. The book *Meaning* is a collection of three

Being raised in a secular Jewish family and living in America during the good old day, Feynman would probably know that the Bible talks about the God of Israel. Nonetheless, the "God" he used in his speeches and writings is more like the impersonal God of modern scientists.

Unlike "new atheists" and ultra-skeptics such as Sommer and Ehrman, Feynman didn't try to claim that the Bible was uninspired and full of errors. From a pragmatic viewpoint, Feynman perhaps regarded Christian ethics as more valuable than the divinity of Christ. He didn't see the Bible as the inerrant word of God either.

But I wonder what Feynman would have to say if he was presented with our evidence. We have shown that 137 can be non-trivially obtained from two special verses in the Bible where pi and *e* are embedded. A scientist must go where the evidence leads.[43]

To obtain three sets of five significant figures each by random chance, the odds are 1 in 90,000 raised to the power of 3, which is 1 in 729 trillion. This probability is two hundred million times smaller and more stringent than the already draconian threshold used by physicists. The nonlinear distance from the null value of "no signal" increases to 7.98 sigma.

To wrap our mind around 1 in 729 trillion, suppose we have 729 trillion dimes. By calculation, they will cover the entire surface of North Dakota, United States. Alternatively, they will cover with more than two layers over Hokkaido, Japan. What is the chance of picking up one specially marked dime in North Dakota or Hokkaido?

If we go by an event per day, this extremely low probability for the triplet of pi, *e*, and alpha to pop out randomly is once in two trillion years. This estimate is much longer than the Big Bang age of the universe of purportedly fourteen billion years.

Physicists take 5 sigma as unwavering evidence for the discovery of a new particle. Logically, physicists can't deny this triplet signal of 7.98 sigma. Our data analysis shows that two verses in the Bible are very strange. The triplet signal can't be explained by random fluctuation. The most natural explanation is divine inspiration.

2.4.4 Implications

Finding the triplet—pi, *e*, and α—from Genesis 1:1 and John 1:1 is indeed surprising. These two verses are unique. They speak about the beginning,

---

previously unpublished public lectures he gave in 1963. By then, he was forty five years old and married to his third wife for the rest of his life.

43. Loeb, "Scientist," title.

Jesus (the *logos*), God, and God's creation of time and the universe. The triplet is like an artist's hidden signature in a masterpiece. It is like an autograph for His Book, the Bible.

The three constants are indispensable in describing the created universe and everything in it. For example, Heisenberg's uncertainty principle requires pi. The Bohr radius of hydrogen atom is dependent on π and the fine structure constant α. As alluded to earlier, the normal probability density function depicted in Figure 2.1, which is central in statistics, depends on π and e.

Furthermore, the fine structure constant alpha is finely tuned. If it were just a few percent different, our universe would be very different and most likely life as we know it wouldn't be possible. So God didn't just create the universe, He also wrote the laws of science for scientists to discover. God knows what value to set for alpha.

Our data analysis clearly shows that these three constants are consistent with the message of Genesis 1:1 and John 1:1. Therefore, we have a piece of objective evidence to affirm a truth about the origin of time and the origin of the universe. These two divinely inspired verses reveal to everyone that Jesus is God, the Creator of the beginning and everything, including the laws of science.

2.5 Is the Entire Bible Divinely Inspired?

Now that we have concrete evidence from our data analysis of Genesis 1:1 and John 1:1, what can we say about the rest of the Bible?

In a letter to Timothy, the apostle Paul wrote about his view of the Scripture, which corresponds to Christians' Old Testament.

> All Scripture is inspired by God [God-breathed] and profitable for teaching, for reproof, for correction, for training in righteousness; so that the man of God may be adequate, equipped for every good work.[44]

And the apostle Peter made it very clear.

> But know this first of all, that no prophecy of Scripture is a matter of one's own interpretation, for no prophecy was ever made by an act of human will, but men moved by the Holy Spirit spoke from God.[45]

44. 2 Timothy 3:16–17.
45. 2 Peter 1:20–21.

Most importantly, what did Jesus say about the Old Testament? Jesus unequivocally endorsed the Scripture as divinely inspired. For example, on the road to Emmaus, ". . . beginning with Moses and with all the prophets, He explained to them [Cleopas and his companion] the things concerning Himself in all the Scriptures."[46]

Referring to the Old Testament, Jesus also said, "For truly I say to you, until heaven and earth pass away, not the smallest letter or stroke shall pass from the Law until all is accomplished."[47]

Responding to the accusation of blasphemy against Him, Jesus said,

> "Has it not been written in your Law, 'I SAID, YOU ARE GODS?' If he called them gods, to whom the word of God came and the Scripture cannot be broken."[48]

In this verse, Jesus quoted Psalm 82:6. The context of this psalm is that the psalmist Asaph prophesized against the magistrate rulers/judges, for they failed to uphold justice. God the supreme Judge called them "gods" but they "will die like men."[49]

The "Law" in this verse is "the Scripture." The "word" here is λόγος (*logos*). It could very well be that Jesus identified Himself as the Word of God (who came to His judging accusers). With that concise argument, Jesus brilliantly asserted that He was unjustly charged with blasphemy. Finally, "the Scripture cannot be broken" is an idiom meaning the Scripture is absolutely true.

Several other passages in the New Testament also show how Jesus regarded the entire Old Testament as having the divine words of God. Our data analysis is consistent with Jesus the *Logos* being God and the Creator. Since God cannot lie, and since Jesus is God, we infer that what He said about the Old Testament must be true. It follows that the entire Old Testament is divinely inspired.

But what about the New Testament?

Since Jesus is God, it goes without saying that all the words He spoke in the four gospel accounts (Matthew, Mark, Luke, and John) and in the other New Testament books were divinely inspired.

Apostle John in his gospel told us how he could remember the words spoken by Jesus. It was none other than Jesus's promise to His disciples:

46. Luke 24:27.
47. Matthew 5:18.
48. John 10:34–35.
49. Psalm 82:7.

> But the Helper, the Holy Spirit, whom the Father will send in My name, He will teach you all things, and bring to your remembrance all that I said to you.[50]

The "you" in this passage is plural. The promise is not only to John but also to the other disciples. This verse is telling us that the Holy Spirit is *the source* that inspires the writers of New Testament. At least John could remember what Jesus had said several decades ago.

As recorded in Acts—a historical narrative of the early church—Paul had a close encounter with the resurrected Jesus. During the apostolic age, the Holy Spirit worked mightily. In Paul's letter to the Corinthians, we read that

> . . . my message and my preaching were not in persuasive words . . . but in demonstration of the Spirit and of power . . . For to us God revealed through the Spirit; for the Spirit searches all things, even the depths of God . . . Now we have received, not the spirit of the world, but the Spirit who is from God, so that we may know the things freely given to us by God . . . not in words taught by human wisdom, but in those taught by the Spirit, combining spiritual thoughts with spiritual words.[51]

Unless evidence suggests otherwise, there is no reason to speculate that Paul was lying. Since Paul said that his message was "taught by the Spirit," at least the New Testament epistles written by him are divinely inspired.

What about the epistles written by Peter, John, James, and Jude? Notice the phrase "For to us" in the above quotation of Paul's letter. It is very likely that "us" might include the authors of gospel and these four writers as well. The reason is that Peter and John were two of Jesus's closest disciples. James and Jude were His half-brothers. They all lived with Jesus for a number of years, and they too should have received the words "taught by the Spirit." Finally, Luke[52] and Mark were close associates of, respectively, Paul and Peter.

The last book in the Bible written by John is of course divinely inspired, since it begins with the "Revelation of Jesus Christ." Again, unless evidence suggests otherwise, there is no reason to doubt John's recording of the visions he saw.

In the history of the early church, religious Jews and the governing Romans persecuted Jewish Christians. The motivation for authoring a book

50. John 14:26.

51. 1 Corinthians 2:4–13.

52. The authorship of *Hebrews* is unknown. I surmise that it is a joint work of Paul and Luke, two highly learned men who wrote in a style different from their other New Testament books.

about things that didn't happen or weren't true in such hostile environment is logically inconceivable. Also, it is a sin to lie and to give a false testimony. As Jesus said,

> "But for the cowardly and unbelieving and abominable and murderers and immoral persons and sorcerers and idolaters and all liars, their part will be in the lake that burns with fire and brimstone, which is the second death."[53]

Critics and scholars of Bible can say whatever they want. But no great scholar, philosopher, or scientist is greater than Jesus—God, the Creator and owner of our universe. Just like the greatest commentator of Shakespearean literature can't be greater than Shakespeare, neither Ehrman is greater than Paul, nor Einstein was wiser than Jesus.

Accordingly, it makes more sense for Christians to listen to what Jesus and Paul said about the Bible than to the opinions of mortal men. They are not responsible for your salvation.

In fact, we can catch a glimpse of the response of early Christians to Paul's teaching. To them Paul wrote,

> For this reason we also constantly thank God that when you received the word of God which you heard from us, you accepted it not as the word of men, but for what it really is, the word of God, which also performs its work in you who believe.[54]

The practice of taking the word from Paul and other apostles as the word of God started as early as the first century. It was only a few decades after the ascension of Jesus.

Therefore, we have a chain of reliable statements to establish that the New Testament contains the divine words of God as well.

Being the Creator of everything from nothing outside of the yet-to-exist spacetime, logically, Jesus is God. What Jesus had promised to His disciples and believers cannot be broken, for God cannot lie. It is highly plausible that the writers of New Testament wrote God's words under the inspiration of the Holy Spirit.

Jesus is also the redeemer of mankind. He came as the Son of man and the Son of God to fulfill the Scripture as the suffering servant,[55] so that "whoever believes in Him shall not perish, but have eternal life."[56] This is the gospel (good news) that Christians embrace and profess as truth. By the joy

53. Revelation 21:8.
54. 1 Thessalonians 2:13.
55. Isaiah 53:1–12.
56. John 3:16.

of receiving the grace of God for forgiveness, Christians are willing to share this good news with anyone willing to listen.

2.6 Conclusion

Christians are often criticized for not having any "scientific" proof of divine inspiration.

But here, it is indeed intriguing that from two related verses beginning with "In the beginning," we obtain pi and the Euler number *e*. These two constants of math are unique in that they are the most important constants in science too.

Just by these two constants alone, which are accurate at the same significant figure, the probability of getting both "signatures of God" by random chance is already 1 in 8.1 billion, which is more than 6.4 sigma. In the jargon of physicists, this doublet signal is counted as a scientific discovery.

The surprise of obtaining the fine structure constant through the sums of the characteristic values of the twin verses brings the probability of being a numerical illusion down to the realm of impossibility. The discovery of this enormously important triplet points to Jesus as the Creator, who said that the entire Old Testament contains the words of God.

The apostle Paul tells us that God reveals Himself evidently to all who are "ungodly and unrighteous." They "suppress the truth in unrighteousness." Since the creation, God made it evident about "His invisible attributes, His eternal power, and His divine nature"[57] through the created world. It has all the birthmarks of design and purpose. This chapter has established a consistent connection between the general revelation in Nature with the special revelation in the Bible.

It is also interesting that the three constants are all accurate exactly at the same significant figures, which is very unlikely to happen if randomness is the driver. Can anyone marginalize these consistent findings as mere random coincidences?

If we doubt the statistical significance of this data analysis, we have to also doubt the discovery of the Higgs particle based on the threshold of 5 sigma. But no physicist doubts that the elusive Higgs particle has already been found, based on *only* 1 in 3.5 million chance. If we don't doubt the discovery of the Higgs particle, logically, we shouldn't doubt this conclusion based on a much more impossible probability of 1 in 729 trillion.

Most likely, the two authors of Genesis 1:1 and John 1:1 were unaware of the fine structure constant α, Euler's number *e*, and possibly also

57. Romans 1.

pi. Moses and John didn't conscientiously embed the three constants into two verses about creation and the Logos. In light of their unawareness, the only viable explanation for this scientific discovery is that a super-intelligent mind, which Christians call the Holy Spirit, was inspiring Moses and John to "push His pencil" with the exact words and spelling.

Many articles and books have been written on the evidence of divine inspiration. This quantitative approach complements their arguments. Although we can argue with formal philosophical devices, nobody can argue against numbers and basic math. Does the Bible have the divine signatures of the Creator? Given the evidence with unusual statistical significance, the logical answer should be "Yes."

3. How is Science Today?

> To me it seems that all sciences are vain and full of errors that are not born of Experience, mother of all certainty, and that are not tested by Experience.[1]
>
> —Leonardo da Vinci

3.1 Power of Science

"How is science today?" is a strange question. It's like asking, "How are you?" A frequent reply is "I am fine." But really? Is the science of historical origins fine?

For a start, I need to define what science is. Rather than giving you the dictionary definition of science, I would like to tell a story instead about my own adventures in science.[2] This narrative approach hopefully will give you a more relatable explanation of what science is and isn't.

I was raised in a non-Christian migrant family before Singapore became an independent nation. My parents were born in China. They didn't have the opportunity to go through many years of formal education. The big questions about the origins are irrelevant when immediate needs demand highest priorities. They were however neither overly superstitious nor overly religious. From young, I was an atheist.

My parents would sacrifice their own little enjoyment to ensure that their children have better education. In school, we learned physics, chemistry, and biology from the textbooks. Using the three laws of Newton, we learned how to solve problems that involved the concepts of mass, momentum, velocity, and force. The reality of these concepts is intuitively experienced in our daily life. Everyone knows intuitively that a fast moving car has a higher momentum than the slower car of the same make.

1. da Vinci, *Notebooks*, 6.

2. Some critics might retort, "Who cares about a small-time scientist from a tiny nation?" I humbly take it that this is *factually* true.

We can't see force and energy but we know they are real because we can *experience* them. We can measure them.

We also learned the periodic table of atomic elements. Though we can't see or smell oxygen, nobody will doubt its gaseous existence. We can see its effect. I remember in an experiment how easily a fire could start with a tiny amount of heat, when oxygen from a small gas tank was supplied to a beaker that contained moist pieces of paper. From helium balloon, lithium battery, neon light, aluminum foil, iron bar, platinum ring, to gold chain, chemistry is in our daily life.

For the rich, possession of a prized diamond gives unmistakable statement about the status of its owner. But diamond is no more and no less than a crystal of pure carbon. This is operational science—repeatably verifiable knowledge about the details of matter, how things interact, and how Nature operates.

Pragmatists may say, "So what?" What is the point of knowing diamond is carbon crystal? Indeed, it won't make you any money by knowing this fact.

Why then do most schools in the secular education system teach science? Once upon a time in history, imperialistic European nations were much more advanced in science and technology. Their advanced military weapons enabled them to conquer and colonize many nations in Africa, Asia, Australia, and America. The colonizers became very rich by taking the natural resources from their colonies. Oftentimes, colonizers resorted to using force to subdue.

China, in particular, was easily defeated by the British Empire, America, France, Germany, Russia, Italy, Austria, and Japan. In the nineteenth century, Japan had the foresight to learn as much as possible from the West. Japan rapidly modernized its economy and became a powerful nation. By the early twentieth century, the Imperial Japan took control of the Korean peninsula, Taiwan, and the North-Eastern region of China.

So, "knowledge is power" isn't a mere cliché. OECD conducted a PISA study in 2018. The sample was seventy eight education systems.[3] Excluding the exceptional China, the top spots went to Singapore, Estonia, and Japan. Despite not having much natural resources, these three countries are relatively rich. Few people can deny the benefits of providing good science education. It empowers us to create wealth.

To illustrate the power of science further, allow me to continue my journey in natural science. At the end of high school (commonly known as junior college in Singapore), as any high school student did, I sat for the

3. Schleicher, "PISA 2018," 8.

Cambridge GCE A Level Examination. My exam results were good enough to earn a scholarship to study in Japan.

But by law, I had to go for National Service first. At that time, they put most of the medically fit students who had already secured an overseas scholarship in Foxtrot company of four platoons. They also mixed us with more street-smart youths.

We went through three months of basic military training. We also learnt to throw a grenade with a proper procedure. Firing an M16 assault rifle at the range was always difficult for me. Nevertheless, most of us were promoted to the Officer Cadet School, where we learned how to fire a machine gun, a mortar, and an anti-tank 84-millimeter howitzer.

Little did I realize then that these weapons have a common ingredient: the explosives invented by Alfred Nobel, a chemist, and a successful businessman. Again, "science is power" literally, and in the case of Nobel, wealth as well.

I graduated from the Officer Cadet School and was commissioned as an officer. For about a month I attended an additional infantry officer conversion course. Like many of the other new officers, I was assigned to a battalion to serve as one of its platoon commanders. Some of my fellow officer cadets had already received permission to "disrupt" from National Service. In September 1980 they left for Cambridge or Oxford, the two most popular destinations for scholarship winners at that time.

In the case of Japan, however, the academic term started in April. So I had to serve until March 1981. The first three months of 1981 were a long wait. I felt a sense of freedom when I was finally released from National Service on the basis of disruption for overseas study.

Under the terms and conditions of the scholarship, I must study mechanical engineering after a year of intensive training in Japanese at a language school affiliated to the Tokyo University of Foreign Studies. The teachers there were experts in teaching Japanese to foreign students, without using English at all. Not only did we learn Japanese, we also studied math and science in Japanese.

By then, I was already a Christian. I went to church regularly, and I started reading Bible frequently. Since I didn't take biology during high school, I wasn't exposed as much to Darwinian evolution. In my memory, the churches I had ever attended in Singapore and Japan, didn't give sermons on Genesis 1–11. I had little idea how big God is.

At that time, I was more absorbed in study than to think about the difficult passages in the Bible. I would rather spend more time reading a few textbooks on physics, trying to figure out how one equation leads to another, and another.

Why study physics on my own? I still can remember vividly that my interest in physics was piqued by a question the physics teacher in the Japanese Language School posed to the class. She asked, "Newton's second law says that the acceleration of an object is directly proportional to the force acting on it. Acceleration is the change of velocity over time. If there is no force, then there is no change in velocity. So the object will be either at rest or move at a constant velocity, which is Newton's first law. Why then is the first law needed when it can be derived from the second law?"

I was stunned by this question. I had never thought about this kind of question that touches the foundation of physics. As any ordinary youth, it never came across to me to question physics textbooks in this thought-provoking way. It was an eye-opening experience.

I became even more interested in physics after reading in a Japanese newspaper about the death of Hideki Yukawa and his works. It was a warm September day in 1981. On the front page, I read a lengthy biographic description of Yukawa and his research in particle physics. This is a branch of physics that tackles a big question. "What is the fundamental constituent of matter?"

In 1949, Yukawa became the first Japanese to be awarded the Nobel Prize. It was barely four years after Japan surrendered unconditionally to the Allies led by General Douglas MacArthur. Atomic bombs had been deployed to destroy the cities of Hiroshima and Nagasaki. The survivors of atomic bombs literally went through hell.

The war with America, which Japan started in the Pacific theater, ended with terrifying affliction and unbearable humiliation to Japanese. For a country struggling to rebuild after the war, the first Nobel Prize for Japan boosted the morale. It was an irony, however, that Yukawa's prize-winning research was on nuclear energy, which brought the angels of death to the two cities—the first in the tragic history of mankind destroyed by atomic bombs.[4]

Such is the devastating power of science. It enabled some European countries, Russia, America, and Japan to create a sense of racial supremacy over the peoples of Africa, Asia, South America, and Australia, lording over them as their masters.

4. For a vacation in September 2020. I went to Ohkunoshima, a small island off the coast of Hiroshima Prefecture. To give you a balanced perspective, I saw the dilapidated facilities and buildings where sarin was once manufactured and stored. The Ohkunoshima Poison Gas Museum preserves the records and used equipment of chemical warfare. I should mention that Japan was an aggressor in World War II, invading several Asian countries with atrocities and war crimes.

But the Bible tells us clearly that there is only one human race, not multiple races. When men don't regard other men of different ethnicity as equally created in the image of God went to war, atrocity increases exponentially with the progress in science and technology. Wars are horrible. Wars brought the Holocaust of man-made hell to Jews. Survivors of Nanjing massacre and nuclear explosions still can recall the hell they went through. There is no winner—even for countries that are highly advanced in science.

In case you may think that I am against science, or Christians are against science, I need to articulate clearly that Christians in general are never against science. Christians are against misusing the power enabled by science. In itself, science is amoral. But it provides the possibility. Nuclear physics was intensely researched in the twentieth century by Yukawa and other scientists before and after him. Prior to the discovery of nuclear radiation, nuclear weapons were not on the table for the generals to even think about it.

3.2 Science at its Best

I entered the University of Tokyo on the basis of admission quota for foreign students in the spring of 1982, when pinkish white cherry blossoms could be spotted at the Komaba Campus. At that time, the campus was hived with activities. Every day without fail, it was filled with cacophony. Some students put up big boards with writings to impeach certain political leaders, and they used loud-hailers to shout that so and so in politics did something bad and must be impeached.

Some of the people were handing out invitation leaflets. They were members of the Unification Church. Many people saw this organization as a cult. I was approached several times because I was usually alone. But I declined and avoided them, as I wanted to make time to study physics.

3.2.1 Anti-Matter

Rather than joining an extra-curricular club, I wanted to learn more about particle physics. This branch of physics describes and explains the quantum behaviors of sub-atomic particles such as electrons and protons. It is also known as high energy physics. Electrons or protons are given such high energy that they are accelerated to a very high speed. The trajectories of these particles in the accelerator are precisely tuned in such a way that they will collide head-on.

I was especially fascinated by how a pure theory could predict the existence of sub-atomic particles before they were discovered. A truly amazing one is Dirac's theory. Published in 1928, Dirac managed to combine quantum mechanics with Einstein's theory of special relativity.[5]

Dirac's theory led to the prediction of anti-particles. For an electron, its anti-particle is a positron, and vice versa. Both particles have exactly the same mass and same quantum spin.[6] The only difference is the electric charge. The electron is negative, while the positron is positive. This anti-electron was discovered in 1932 with an apparatus called the cloud chamber. Charles Wilson invented this observation-enabling instrument. We can detect anti-particles over and over again. Anti-matter has become an operational science.

On a family vacation trip to Japan in 2015, we visited the National Museum of Nature and Science in Ueno, Tokyo. An underground exhibit of cloud chamber was there. I saw many mist-like trails appearing and disappearing in it. These trails were caused by the charged particles of cosmic rays, much like the white contrails caused by airplanes high in the clear blue sky. Anything that begins to exist has its cause. The trails that begin to appear have their cause in cosmic rays.

This is science at its best. Dirac's elegant theory produced a *speculation* about the existence of anti-matter. If you think about it, anti-matter is a pretty crazy idea. If nobody had found an anti-particle, Dirac's theory would become a questionable speculation. Nevertheless, Dirac's theory was proven over and over again by experience. Based on this operational science, the technicians of a national museum constructed an exhibit to let the public "see" the sub-atomic particles of cosmic rays. They include Dirac's positrons and pions.

Incidentally, pion is the particle predicted by Yukawa's theory.[7] The experimental verifiability and repeatability of these theoretical speculations constitute the backbone of scientific research. They have become science. It is also a vindication that quantum mechanics is how Nature really operates at the sub-atomic level. Theories of Dirac and Yukawa are operational science.

Have you ever heard of PET scans? Today, we know that certain radioactive materials emit positrons. The Positron Emission Tomography (PET)

5. Dirac, "Quantum," 619–624.

6. The spin of an electron was introduced to describe one of its physical properties. This quantum characteristic is analogous to the classical notion of rapid rotation. For its ability to account for the spectral lines, see Uhlenbeck and Goudsmit, "Spinning," 264–265.

7. Yukawa, "Interaction," 48–57.

scan is a medical diagnosis tool to detect brain tumors etc. Science at its best often leads to beneficial applications.

3.2.2 Quantum Physics

I managed to convince my supervising professor Dr. Naomasa Nakajima that quantum mechanics would be important even in mechanical engineering. He approved my senior-year thesis on the simulation of a quantum particle colliding with a quantum barrier.

It must have been a bizarre topic because Professor Nakajima's research was in the field of mechanical engineering. His expertise was in the area of industrial robotics. Although I could read and write in Japanese, he allowed me to write my graduation thesis in English. I also received help from a fellow student Takashi Hasegawa. He developed a PCB interface with the printer. It was a kind of mini robot at that time. I used it to print the simulation pictures generated by a computer program that I wrote in C language.

I solved the Schrödinger equation of quantum mechanics for one- and two-dimensional spaces numerically by computer programming. The Schrödinger equation is the most fundamental equation of quantum physics. It is the sub-atomic analog of the classical Newton equation of motion.

For each simulation, I constructed a bell-shape packet of wave-matter and gave it a speed to move toward a quantum barrier.[8] When it hits the barrier, my simulation results show that there is a reflection. This rebound is what you expect when you throw a tennis ball at a wall. But the strange thing about quantum mechanics is that the rebound is only a portion of the original packet. Certain portion is trapped in the quantum barrier under some conditions, and certain portion goes through it. This transmission of a quantum particle through a barrier is crucial. Without it, microscopy with electrons will be a fantasy.

Engineers make use of this weird quantum property of transmission to construct a microscope. It is thus a nice idea that "transmission" is attached to "electron microscope" and we have the acronym TEM in the field of cryo-electron microscopy. It turns out that I had a chance to operate a TEM later in my pursuit of a master's degree by research.

This is another example of science at its best. In the Schrödinger framework, transmission of a beam of electrons through a barrier—the specimen to be magnified—is predicted. TEM is a working testimony of this scientific fact. For simple quantum barriers, we can even compute the percentage of transmission analytically.

8. Ting, "Quantum," 1–146.

But Nature isn't so simple. Most of the practical problems involve quantum barriers for which analytical solutions are unobtainable. We can nevertheless solve the Schrödinger equation numerically. We can simulate the dynamics of a quantum system. This research approach is known as computational science.

Most importantly, science at its best tells us its limits. In quantum physics, the Heisenberg principle states that it is theoretically impossible to measure both the location and momentum of a particle precisely at the same time. Separately, you can measure either of them as precisely as you want. Simultaneously, however, both can't be measured precisely.

Heisenberg's uncertainty principle has its root in the quantum wave-particle duality. Every electron has as if a dual personality. At the same time it is a particle and a wave.

Intuitively, we find it difficult to visualize a spatially extended wave as a spatially localized particle. We could imagine, nonetheless, the ripples in a still lake when we gently drop a pebble into it. The remarkable catch is that, in quantum physics, the pebble doesn't cause the ripples to happen. The pebble and the ripples *are* one entity. Perhaps this wave-particle duality gives us a glimpse of the nature of Jesus being both the Son of Man and the Son of God at the same time?

Like anti-matter, wave-particle duality sounds highly speculative. But it is experimentally proven to be true, over and over again. Wave-particle duality isn't speculation anymore. It has become science. This is how Nature operates at the quantum level. Even so, how exactly does the quantum world morph into our familiar classical realm? This problem remains an unsolved holy grail of physics.

By any estimation, wave-particle duality is one of the most exotic aspects of quantum mechanics. A well-known quote attributed to a distinguished physicist Richard Feynman goes like this. "Anyone who claims to understand quantum theory is either lying or crazy."[9] The same can be said of the Holy Trinity—the Father, the Son, and the Helper.[10] Three persons but one God.

In quantum physics, we also have the energy-time uncertainty principle. If we measure the time of a quantum state very precisely, then we can't measure its energy definitely. Conversely, if energy is measured precisely, then we can't tell for how long the quantum state can exist.

9. Moskowitz, "Physicists," line 1–2.

10. John 14:26.

3.3 Real Science and Fake Science

When I was in my fourth year at the Mechanical Engineering Department, I took the entrance exam for the Physics Department of the Graduate School of Science, University of Tokyo. It was at the tail end of 1985 summer. This time round, I had to compete with top physics students for a spot. To my surprise, I qualified. I suspect I barely made it. Nonetheless, I was awarded another scholarship to continue my graduate study in Japan.

3.3.1 Muscle Protein

My first choice was to do particle physics. I went for the interview for the qualified candidates. I can still recall that Professor Tohru Eguchi, a famous theoretical particle physicist, was in the panel. Somehow I was quite nervous, and I didn't impress them.

My second choice was in biophysics, which was not popular at that time. So for the next two years, I did experimental physics on the three-dimensional structure of a protein in the lab spearheaded by Professor Takeyuki Wakabayashi. Meanwhile, I continued my own study in theoretical physics, out of pure interest.

Looking back, the two-year experience of doing experimental works is a blessing in disguise. Most theoreticians tend to simplify matters and underestimate the complexity of reality. I learned what science really is through experiments. This experience enables me to discern between speculative claims of a theory and facts from experiments.

Doing experiments is a lot of hard work. Professor Wakabayashi had developed and set up a system to analyze the three-dimensional structure of a muscle protein. The most difficult part for me was to slaughter a rabbit acquired from a supplier. The purpose was to extract fresh muscle proteins.

Professor Wakabayashi was a medical doctor as well. I followed his medical procedure to inject certain drug through a vein in the rabbit's ear for sedation. Then I used a surgical knife to slit the rabbit's throat. I had to follow the lab procedures very closely; no margin of error was allowed.

Then I quickly took the flesh sample (myofibrils) from the rabbit's back and put it in a sanitized test tube. I processed the sample with a centrifuge at some stage to obtain filaments that contain a muscle protein. Its name is single-headed heavy meromyosin.[11]

11. The lab procedures are much more complicated than this simplified description. See Ting, "Three-Dimensional," 1–69, my master's degree thesis.

Throughout my master's degree program, I killed two rabbits under the supervision of Professor Wakabayashi and under the instruction of Dr. Makio Tokunaga, a hardworking and smart postdoc, and Ms. Kimiko Saeki, a talented lab technician. Our scientific research was to find out how muscle generates force at the molecular level.[12]

From killing a rabbit to low-dosage electron microscopy of the muscle proteins as shown in Figure 3.1, I had to conduct thirty six hours of operations non-stop. It's because proteins decay easily. Any slight deterioration will distort the three-dimensional structure of the protein. When that happened, all the work up to that point of time of making a mistake would be wasted, and I had to start all over again.

Proteins are fragile. When I did my electron microscopy, to minimize the damage to the specimen, the dosage of radiation on the specimen was no more than ten electrons per atom size over 2.8 seconds. The electric current flowing through each atom is about a quintillionth[13] of the current we use to charge our mobile phone with a USB.

From this experience and training in biophysics/molecular biology, I learned that proteins decay rapidly outside of a living cell. In other words, muscle proteins won't decay rapidly in the rabbit; but shortly after they are extracted and shelved even in a special-purpose fridge, the process of decay begins and accelerates. In fact, you don't need to be a scientist to know that raw fish left too long in an ordinary home refrigerator will surely become distasteful.

12. My first scientific paper was published in 1989 in the Journal of Biochemistry (Saeki et al., "New," 606–511.). In the Japanese custom, the last author is usually the research group leader. In our case, it was Professor Wakabayashi, who secured the research funding and provided the research ideas.

13. A quintillionth is 10^{-18}, which is a decimal point followed by seventeen zeros and then one.

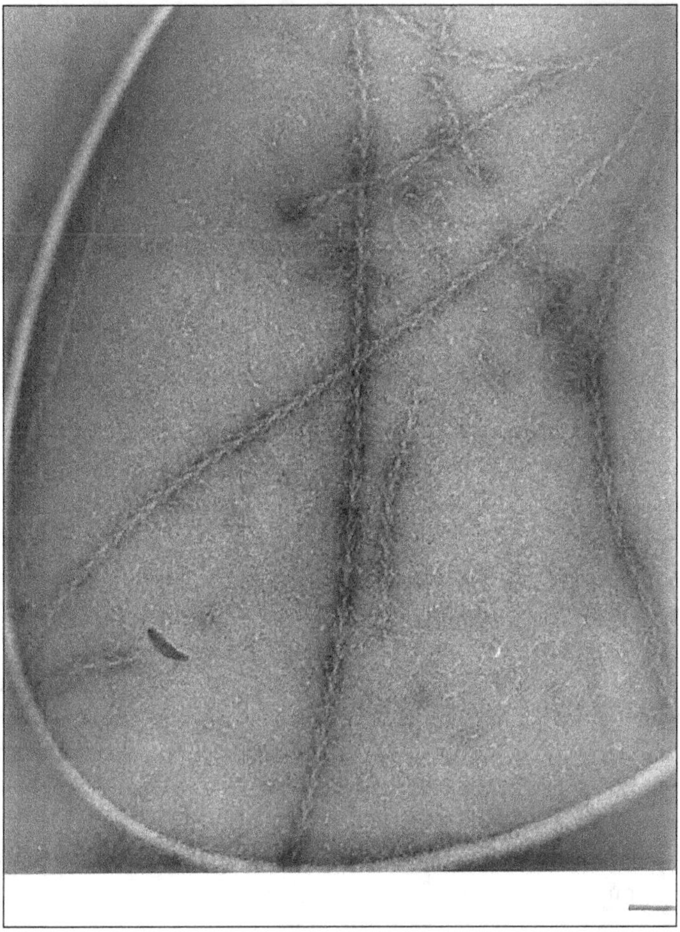

Figure 3.1. This is a low-dosage micrograph of muscle protein filaments I took with a scanning transmission electron microscope (TEM). This micrograph is in my master's degree thesis. The name of the protein complex is actin-tropomyosin-single-headed-meromyosin. The bar represents 100 nanometers. Characteristic "arrowhead" can be seen along the filaments. A particularly long filament has "arrowheads" running vertically from top to bottom.

3.3.2 Darwin's Pond

This every-day, common sense evidence doesn't seem to be taken into consideration by Darwin. In a letter to J. D. Hooker in February 1871, he wrote,

> It is often said that all the conditions for the first production of a living organism are now present, which could ever have been present.—But if (& oh what a big if) we could conceive in some warm little pond with all sorts of ammonia & phosphoric salts,—light, heat, electricity &c present, that a protein compound was chemically formed, ready to undergo still more complex changes, at the present day such matter w^{d} be instantly devoured, or absorbed, which would not have been the case before living creatures were formed.—[14]

In Darwin's own admission, the primordial soup was a wild speculation of the origin of life, as he exclaimed, "& oh what a big if." It just seems so easy for proteins to emerge from an unknown pond.

The world we observe at present clearly suggests that the irreversible origin of life must have occurred in the historical past. Nobody can travel back in time to find some "warm little pond," somewhere no one knows, and some millennium no one can tell. Darwin's hypothesis is non-testable today, as he said himself, "at present, such matter would be instantly devoured, or absorbed." We must go back to unknown time "before living creatures were formed." Observations of the origin of life are simply impossible.

Darwin's speculative "warm little pond" of the origin of life can't be observed at all. By definition, it is unscientific because we can't even find out the chemicals in it, let alone test it repeatedly. Unbiased evidence for life from non-life is nowhere to be found. Darwin's speculation has the appearance of science when it is not, because it is unobservable, untestable, and thus unscientific. We don't know whether Darwin was serious about this wild guess.

By contrast, as trivial as it may seem, suppose we hypothesize that proteins will decay. This hypothesis of protein decay can be tested scientifically. We put a fresh chicken egg in a fully sanitized box. Then we seal the box perfectly. We keep the box at room temperature in a room underground. If, after a year, anyone who knows how to analyze the three-dimensional structures of the egg proteins—which are known—and find that they do disintegrate, then our protein decay hypothesis is proven correct or verified. But if the egg proteins remain structurally intact, then our hypothesis is proven wrong.

Instead of one set of an egg in a box, we can have multiple sets. We carry out the same experiment for a few years. This is a simple illustration of a basic principle of science at work—repeatable verifiability.

14. Darwin Correspondence Project, "Letter no. 7471," accessed on 15 July 2022. https://www.darwinproject.ac.uk/letter/?docId=letters/DCP-LETT-7471.xml.

Interestingly, one way to define verifiability—a key principle of accounting—is as follows. "A company's accounting results are verifiable when they're reproducible, so that, given the same data and assumptions, an independent accountant can produce the same result the company did . . . If not, the result isn't verifiable."[15] Repeatable verifiability is crucial in accounting too.

Science is based on repeatably verifiable evidence. But as any scientist will confess, being proven wrong is the most undesirable outcome in the career of any scientist. To avoid the embarrassment, we can cheat and say, "Let's wait for seven billion years and then open the box to test."

Common sense tells us that given a longer time, everything else being equal, cumulative probability of decay will increase. So if we insist to wait for seven billion years, then our hypothesis can't be tested to be either right or wrong—which is unscientific—for by then our Sun will run out of fuel, and nobody will be around to perform the test. The hypothesis has become a non-testable speculation. It is not science anymore.

By the same token, Darwin's speculation of the historical process of evolution is unscientific. The reason is straightforward. It is no different from the protein decay experiment, except that big bio evolution moves in the opposite direction of time. Instead of everlasting long time to wait forward to the future, it is to wait dark deep time backward to the past. And instead of devolution of protein structure, it is evolution to more complex protein varieties from chemicals in the pond.

Evolutionists must assume that the pond has the required concentrations for synthesis. But any practicing synthetic chemist or biochemist can testify that it is extremely difficult to synthesize a chemical with sufficient yield even in a controlled laboratory.[16] The procedures and timing must be exactly right. If a mistake is made at step seven, everything is undone, and you have to go back to step one and try again. That is, you must go back and gather sufficient quantities of raw materials. You must know the exact ingredient, the exact proportion, the exact timing, as well as concentration, temperature, etc. There are so many details; not an iota margin of error is allowed. How can a mindless pond know what to do?

3.4 Abiogenesis

A modern version of Darwin's pond is abiogenesis. Let's take the following definition of abiogenesis from Encyclopaedia Britannica.

15. Merritt, "Verifiability," para. 2; CFI Team, "Qualitative," sect. 5.

16. See a presentation by James Tour, a leading synthetic chemist (Tour, "Mystery.").

> Abiogenesis is the idea that life arose from nonlife more than 3.5 billion years ago on Earth. Abiogenesis proposes that the first life-forms generated were very simple and through a gradual process became increasingly complex.[17]

In the same Britannica article on abiogenesis, it is said that "[abiogenesis] became impossible once Earth's atmosphere assumed its present composition."[18] ". . . abiogenesis has been neither proved nor disproved."[19]

"There remain many unanswered questions concerning abiogenesis. Experiments have yet to demonstrate the complete transition of inorganic materials to structures like protobionts and protocells and, in the case of the proposed RNA world, have yet to reconcile important differences in mechanisms in the synthesis of purine and pyrimidine bases necessary to form complete RNA nucleotides."[20]

This description of abiogenesis is consistent with the assessment of Klaus Dose, an expert in biochemistry. He wrote,

> More than thirty years of experimentation on the origin of life in the fields of chemical and molecular evolution have led to a better perception of the immensity of the problem of the origin of life on Earth rather than to its solution. At present all discussions on principle theories and experiments in the field either end in stalemate or in a confession of ignorance.[21]

One of Darwin's bulldogs in continental Europe was Ernst Haeckel. He wrote that the origin of life was so hard to crack by scientific methods. Exasperated, he wrote,

> If we do not accept the hypothesis of spontaneous generation, then at this one point of the history of development we must have recourse to the miracle of a supernatural creation.[22]

Abiogenesis—life from non-life material on Earth with unknown processes, unknown mechanisms, and unknown conditions—like Darwin's pond, is not scientifically testable. No one can observe and verify the hypothetical set of various "prebiotic" conditions really did actually happen on

17. Rogers, "Abiogenesis," para. 1.
18. Rogers, "Abiogenesis," para. 1.
19. Rogers, "Abiogenesis," para. 2.
20. Rogers, "Abiogenesis," para. 11.
21. Dose, "Origin," 348.
22. Haeckel, *History*, chapter XIII.

Earth dark deep time ago. No one could wait for billions of years for the first single-cell life to happen from non-living organic materials.

A cell is gazillion times more complex than a bunch of amino acids and ribonucleic acids. A comparison of a cell with a test tube of organic chemicals doesn't make sense, because they belong to different categories. It is like comparing a sheep with Einstein. If you fully understand and *experience* the complexity of a living cell, you will find a deep chasm between a cell and a test tube of biochemicals. The unbridgeable chasm is beyond the wildest imagination.

Is there a theoretical model to compute how long it would take to synthesize a protein? Have scientists ever observed the hypothesized chemical processes *in situ*? Abiogenesis is hardly what I have ever come across as science.

Francis Crick, a co-discoverer of the double helix, was an arch-atheist. He described the situation of abiogenesis as follows:

> An honest man, armed with all the knowledge available to us now, could only state that in some sense, the origin of life appears at the moment to be almost a miracle . . .[23]

Crick conceded that the origin of life is "almost a miracle." But most likely he still believed that big evolutionists would one day explain away this "miracle" in the language of historical science.

Even so, nobody has ever demonstrated by experiments that a muscle protein can be synthesized from organic materials. If Nature were to assemble a single-headed heavy meromyosin from scratch, would four billion years suffice? Do we need to remind big evolutionists that we also need several other proteins *simultaneously* for a muscle fiber to function properly?

Furthermore, it's a fact that a living cell needs DNA to give instructions for assembling every protein in it. Conversely, DNA needs proteins to replicate. You need both to be present at the same time in working conditions. But even if both are present, without a protective cell membrane, proteins and DNA will decay easily.

Living things are incredibly complex. A model as elegant as the Dirac equation is simply impossible.

Therefore, we need discernment. Unscientific hypotheses such as the primordial soup of Darwin in particular and abiogenesis in general are advertised as real science. Observational proof isn't forthcoming; no one can time travel to the past to observe abiogenesis. Suppose scientists said they

23. Crick, *Life*, 88.

found a bowl of primordial soup somewhere. But how could they prove that it is really the *historical* one?

At least two evolutionists recognize this problem of historicity.[24] They highlight two distinct facets of abiogenesis: historic and ahistoric. They conclude that scientists probably will never know the exact *historic* path by which life on Earth emerged. In other words, *historical* abiogenesis can't be known. Yet, on the ahistoric front, they still believe that non-life can become simple life and then complex life.

If scientists could chemically synthesize a living cell from raw materials, they would have a strong case to convince even the die-hard skeptics. But as it stands now, all natural theories of life's origin are untestable speculations. Origin is "almost a miracle." This assessment is consistent with the findings of, for example, Tan and Stadler.[25]

3.5 Dark Deep Time

What is certain at the end of the day is that for whichever new models of evolution since Darwin, the magic wand is always dark deep time. Whenever evolutionists declare "about *x* million years ago," how is it different from "once upon a time?"

To gain some intuitive grasp of dark deep time, let's examine the timeline of human evolution. Timeline is a euphemism for history. The commencement of life was somehow 4.2 billion years ago.[26] Big evolutionists are telling a story about the past. It is history, not science.

Let's take this amount of 4.2 years of natural history of evolutionists as equivalent to an hour before midnight. To wrap our mind around this billion-year time scale, I plot the commencements of new lifeforms in Figure 3.2 as sixty minutes before midnight.

24. Pross and Pascal, "Origin," 120190.
25. Tan and Stadler, *Stairway*, 1–267.
26. https://en.wikipedia.org/wiki/Timeline_of_human_evolution.

Figure 3.2. Relative time scale of purported human evolution.

According to some evolutionists, in the sixty-minute history of big evolution, the Earth had no animal for the first fifty one minutes—no bees, no butterflies for the plants. The Earth waited until nine minutes before midnight for the primeval animal lifeform to appear. Modern humans came to the scene fifteen seconds before midnight.

Big evolutionists claim that Hominidae were the common ancestors of humans, chimpanzees, gorillas, and orangutans today. They are said to have appeared about 17.5 million years ago. Some members of Hominidae—which *were* animals—for unknown reasons, decided that they wanted to walk upright rather than climbing trees. They somehow became humans about 0.055 million years ago.

How could it be that within such a relatively short span of time, a Darwinian *favored race* could feel compelled to make for themselves loin coverings? Isn't this practice of covering the reproductive organs an impediment to reproducing offspring, which is *the* critical factor for survival?

From Figure 3.2, it looks like the process of evolution is accelerating. Over the last fifty five thousand years since humans were said to have first appeared, and given the acceleration of the process of evolution, why don't we have a human with wings? Also, why don't we have amphibian humans? Clearly the additional ability to be in the air or in the water provides a clean competitive advantage in the struggle for survival and dominance. So, why is there no biodiversity for human race?

Can big bio evolution models say that human evolution has stopped? If not, can they predict the anatomy of humans two thousand years in the future?

What about plants? Why can trees of different species coexist in the same land while constantly competing for nutrients and sunlight? Why is there such a wide diversity of plants? Empirical evidence is very clear that plant cells and animal cells are very different. So how did plants become animals?

The idea of big bio evolution raises more questions than the answers it tries to provide.[27] Yet all abiogenesis hypotheses, which inescapably presuppose dark deep time, are marketed with the ringtone of science.

"Don't you know the famous experiment done by Miller in May 1953? Don't you know that many subsequent experiments have shown that organic compounds can be formed?"

Sure. These experiments are scientific and repeatable. That's why the public is confused into believing that big evolution is science. Discernment of the chasm between operational science and historical science is lacking. These experiments prove nothing. You can find amino acids in your kitchen sink after you wash a piece of meat. Do you expect new lifeforms to appear in your kitchen?

There is a fatal assumption of abiogenesis. Oxygen must be *absent* dark deep time ago. Miller's research group reported in 2007 that to increase the yield of a handful amino acids, oxidation inhibitors such as iron must be added.[28] But is it verifiable that such inhibitors existed in the presupposed primordial soup?

Also, it is a mystery that almost all the amino acids of proteins from life are left-handed. The Miller experiments have only a fifty percent chance of getting left-handed amino acids. How do you separate left-handed from right-handed chemicals?

In summary, dark deep time is a big god of the gap in big evolution. Nobody has any living experience of dark deep time required by historical science. Leonardo da Vinci was right. "All sciences are vain and full of errors that are not born of Experience."[29] Should we embrace big bio evolution without a grain of salt?

3.6 Forget about Jesus!

Most of the fundamental laws of physics and chemistry are well established by the twentieth century. Some imperialistic scientists aren't willing to work on mopping up the details at their home soil of present-time operational science. They venture into colonizing the biblical genesis of time, space, matter, life, as well as consciousness, mind, and personality.

Many Christians compartmentalize science as science and religion as religion. There's no overlap, so let's move on to avoid church division. Some bow down before big evolution and change the literal interpretation

27. Guliuzza, *Twenty*, 5–217.

28. Cleaves et al., "Reassessment," 105–115.

29. da Vinci, *Notebooks*, 6.

of Genesis. This is the consequence of mistaking historical science as operational science.

Let's see how we can untangle historical science from operational science. Take for our case study, Lawrence Krauss, a theoretical physicist. During a talk in 2009, he said,

> . . . I wrote that book because it is the most poetic thing I know about the universe . . . and I wrote a book that was about a different ancestor. Still it's called Adam, but the amazing thing is that if every atom in your body came from a star that had exploded and the atoms of your left hand probably came from a different star than your right hand, it really is the most poetic thing I know about physics. You are all star dust. You couldn't be here if stars hadn't exploded because the elements—the carbon nitrogen, oxygen, iron—and all the things that matter for evolution weren't created at the beginning of time. They're created at the nuclear furnaces of stars and the only way they can get into your body is if the stars were kind enough to explode.[30]

To the rousing applause of all the atheists in the audience, Krauss gleefully delivered the punch line. He exclaimed,

> So forget Jesus! The stars died so that you could be here today.

Krauss was saying that you are alive not because of Jesus, the Creator and the Lamb of God. Rather, it was the death of stars that gives life. "You are all star dust." You are made of dead stuff. Big evolution relies on death to progress.

Why are cosmologists fond of saying that we are star dust? First, they tell us that Big Bang produced primarily hydrogen and helium. This part is historical science. We can't observe it. From the periodic table, we know that these two elements are the lightest ones. We also know that stars are made of hydrogen. This part is operational science. We need to differentiate, and it's not difficult.

The question is, how do we account for the heavier elements found on Earth? Cosmologists tell us that these elements are produced by nuclear fusions inside every star. When massive stars burn out, they explode and send the heavier elements they have produced into space as star dust. Star explosions are observation science. We can't deny these facts.

But are we really the star dust of dead stars? Is there evidence that dead stars were the cause of uranium and other heaviest elements we have on Earth?

30. Krauss, "Stars."

In today's culture, despite the evidence of living Jews, doubting the historicity of Adam and Eve is lauded as rational. But just because we have uranium on Earth, doubting the historicity of big evolution is often derided as unreasonable and laughable. So, in order not to be embarrassed, should Christians read Genesis with big evolution as an *add-on* belief ?

But who am I to critique our Christian brothers and sisters who accept big evolution? Not until recently, I was a believer of big cosmic evolution too. I didn't examine the details and I trusted what mainstream physicists and biologists are saying. I even gave a talk about Big Bang in a small church gathering twenty years ago.

3.6.1 Get Rid of Star Dust with Science

But when I started to think carefully, I realized that the scientific basis of big cosmic evolution is dubious. To underscore this point, let's respond to "you are star dust" with arguments based on science and common sense.

First, take Supernova 1987A as an example. This luminous stellar explosion occurred in 1987 about 168,000 light years[31] from Earth, the nearest in recent centuries. Astronomers have estimated an average of one supernova for every half a century in our galaxy.[32]

We can see the light from a supernova. But all atoms run much slower than light. Over a *vast* distance, did the slower heavy atoms from a supernova manage to land safely? If we are star dust, then the *abundance* of atoms heavier than helium on Earth requires *a lot* of debris from every star explosion near us to land here. But this goes against the cosmic Copernican principle, which is actually a philosophical assumption. It basically says that Earth isn't a special place for landing. So Earth should receive only a few debris from each nearby supernova. The account doesn't tally.

Second, it is an observed fact that every supernova lasts for a few months. If the heavy elements found on Earth *originated* from supernovae, then the distribution of say, gold, should be uniform.

As an illustration, suppose you have a ball and you give it a spin. You then spray powder particles that can stick to the spinning ball. It is easy to see that the sticky particles won't concentrate on a few places. You will find that they are almost everywhere on the ball.

In reality, gold isn't found everywhere on Earth except at the gold mines. While South Africa has one of the largest gold mines in the world,

31. Note that "light year" is a measure of distance and not time. It assumes one-way speed of light is a fixed constant.

32. Naeye, "Milky," para. 1.

Western Europe has almost none. Therefore, mines of precious metals blast the star dust speculation away. It is a myth propagated by people who consider themselves as "science guys." They are supposed to follow the evidence wherever it leads.

Third, high-energy radiation from supernovae is harmful to life rather than life giving. You don't need to be a scientist to know that ultraviolet (UV) radiation from the Sun is harmful to your skin.

To test this claim, we just need to spend a few hours at the beach. Without a shade and without applying sunscreen, we will suffer from sunburn. Now, star explosions produce deadly potent gamma rays, which are about 13,000 times more energetic than UV radiation. We all know the damaging power of UV light. How much more deleterious if a cell is radiated by gamma rays? As an illustration, if it takes four hours to get sunburn from the UV light, under gamma radiation, about one second will be enough to kill the skin cells.

Finally, where on earth is the empirical evidence for the claim that every atom in your body came from an exploded star? How is it possible to test that atom *x* of your right hand is from star *y*, and atom *u* of your left hand is from star *v*? These stars had supposedly "died" dark deep time ago. How then could any scientist provide evidence to support such a claim?

The popular chant that "we are made of star-stuff"[33] is not testable. By misusing observations of astronomy to speak on matter concerning the *history* of Earth and stars, scientists aren't scientists anymore but your chatty uncles next door.

3.6.2 Star Dust Contradicts Genesis

Why should Christians take issue with the "poetic" star dust? Isn't it quite benign to believe that the origin of the elements on Earth is a bunch of unknown dead stars?

No. It is in fact malignant because this star dust claim is inconsistent with the biblical account.

First, it means that the Earth wasn't created at the beginning but billions of years after the Big Bang, the supposed beginning of time. It means that stars were created before the Earth. But Genesis 1 reveals that Earth was created first. In a nonchalant way, God later made "the stars also."[34]

Second, the claim that heavy metals in big stars eventually came to Earth by supernovae is inconsistent with a biblical passage in Genesis 2. We

33. Sagan, *Cosmic*, 190.

34. Genesis 1:16.

read that the "land of Havilah" has gold, and "the gold of that land is good."[35] We can infer that when God created the Earth, He had already made gold. It is therefore reasonable to think that He provided all the elements on Earth in the beginning. Carbon, nitrogen, oxygen, iron and all things that matter to our physical body didn't come from outer space.

Third, atheists' belief that we are made of star dust directly conflicts with human creation in the Bible. On the sixth day, God "formed man of dust from the *ground* . . ."[36] Also, God told Adam after he sinned, "By the sweat of your face you will eat bread, till you return to the *ground*, because from it you were taken; for you are *dust*, and to *dust* you shall return."[37] The Bible tells us that we are earth dust, not star dust.

In conclusion, "star dust makes us" is not science because it isn't testable. I don't deny that elements heavier than helium are being made in stars right now. We can observe them with spectroscopy. This part is observation science. But to say that they came to Earth dark deep time ago is another matter altogether.

Also, just because the heavier elements were produced by stars doesn't necessarily mean that they had reached us in the supposedly distant past. No cosmologist can observe uranium atoms traveling across many light years of distance to Earth. Cosmologists can't test that the gold chain you own came from stars that don't exist anymore, can they?

If Christians accept star dust, which has no scientific basis, then Genesis 2 has to be interpreted very differently from its plain meanings. Eventually, Christians may have to give up God created the Earth, or to cook a convoluted non-literal interpretation. Either way, atheists will still point out that the Bible is wrong about the origin of gold. But "the land of Havilah" is more consistent with the fact that gold isn't distributed evenly on Earth. The star dust claim is completely unscientific and unbiblical.

3.7 Dogmatism of Big Evolution

Big bio evolution is an excessive extrapolation from the fact of adaptive mutations to the unobserved transmutations of species. Evolutionists argue that it *was* possible for simple organisms to evolve into complex lifeforms, and then into humans. They believe that Nature favors the emergence of new beneficial traits. But as evolutionists always say, for a transmutation to finally occur, it takes millions of years.

35. Genesis 2:12.
36. Genesis 2:7.
37. Genesis 3:19.

3.7.1 Reasons to Doubt Darwin

Today, we know that genetic mutations within a population of fruit flies are scientific facts. But to suggest that fruit flies are descendants from some unobservable common ancestors requires belief.

Also, thanks to the short life span and high reproduction rate, scientists have thousands upon thousands generations of fruit flies. Yet, we have never heard the news of fruit flies evolving into, say, dragonflies, or other more complex insects.

The scientist-guided unnatural selection of fruit flies by some "desirable" traits in a closely controlled environment can't produce a dragonfly. Do you expect unguided natural selection in an open milieu can do better?

Similarly, due to genetic mutation, bacteria such as E. Coli change over time. But what is important to note is the fact that E. Coli, after countless generations experimented in the laboratories worldwide, remain as bacteria. They don't and can't evolve into some other unknown entities of higher lifeform such as moss. Incidentally, geneticists have found that moss has ten thousand more genes than us.[38] Simple moss plants outperform us by gene count. For having more genetic information, is moss more complex than us?

From the standpoint of information science, new information can't be obtained without *deliberate* editing of existing information. This fact is obvious to anyone who has ever written an essay. Without more inputs, its information content will remain unchanged.

Now, DNA inside a cell functions as a biomolecular storage of genetic information. For example, there is no DNA in human genome to code the information for the development of two large wings that allow humans to fly. Flying like a bird is a desirable trait. But we aren't designed for that. It's not in our DNA.

To illustrate this fact, suppose this page contains three hundred words. Through biotechnology, the letters used in this page are encoded as DNA sequences. No one will believe that these DNA sequences would, after billions of years, become DNA sequences of different makeups that, upon decoding, a masterpiece of Shakespeare would appear. The simplest reason is that the number of words in this page is much less than Shakespeare's shortest play, *The Comedy of Errors*, which has about fourteen hundred words. Where do the extra eleven hundred words come from?

Next, let us consider the reverse situation. We encode the letters used in *The Comedy of Errors* as DNA strands. This time round, we have more words than this page has. It is now possible, by chance, produce the content

38. Zimmer et al., "Reannotation," 498.

of this page in the exact order of the words. But eleven hundred words have to be discarded. We have a loss of information.

More seriously, we have a stopping time problem. How does Nature know when to stop the process after it has produced this page by accident? To stop, Nature must know the content in this page *before* the process begins. Do you see the problem?

In the same way, Nature must know what a bird *is* before starting the evolution process to change dinosaurs. Isn't it cheating?

Big evolution by *gradual* accumulation of microevolutions is questionable. Some scientists have concluded that this theory is fundamentally wrong. It is not supported by observation data. For example, a leading evolutionist, Stephen Jay Gould, described the scientific evidence as follows. "The fossil record does not support it; mass extinction and abrupt origination reign . . ."[39] He continued,

> To sidestep this unpleasant truth, paleontologists have relied on the extreme inadequacy of the fossil record—all the intermediate stages are missing in a record that preserves only a few words of the few lines of the few pages left in our geological book.[40]

If the intermediate stages of evolution are absent in the fossil record, then Darwinian evolutionists have little evidence to assert that gradual big evolution is a scientific fact.

To "sidestep" "the extreme inadequacy of the fossil record" in Darwinian evolution, Gould and his collaborators proposed a new idea, which they called "punctuated equilibrium." The "equilibrium" is a long period of tranquility called stasis—a euphemistic jargon for "no macroevolution." The adjective "punctuated" refers to "intermittent bursts of activity and volatility in which many species become extinct and new ones emerge."[41]

This modern speculation of Gould is more consistent with the fossil records of stasis—little change. Viewed from another angle, it is probably not wrong to say that Gould's punctuated equilibrium is a fudge to rescue big bio evolution from the failure of gradual Darwinian evolution.[42]

From the biblical standpoint, there was only one "puncture" in six days. On day three, plants were created. All animals were created in the last two days.

39. Gould, *Ever*, epilogue.

40. Gould, *Ever*, epilogue.

41. Bak, "How," 29.

42. To know more about post-Darwin evolution theories, I recommend Meyer's *Darwin Doubt*. (Meyer, *Darwin*.)

3.7.2 Darwinism is a Secular Religion

In spite of the new developments in big bio evolution such as Gould's work, most textbooks still state that Darwinian macroevolution is a scientifically proven theory. Is there something amiss?

Darwinism has become an ideological orthodoxy. Look, nobody celebrates "Mendel Day" or "Newton Day" *every year.* But "Darwin Day" is observed annually by humanists/atheists to celebrate the birth of Darwin, to promote the teaching of evolution and Darwinism in schools, and to educate the general public about Darwinian evolution.[43]

Isn't it quite strange that Darwinian evolution needs promotion? For the operational science, nobody will find it necessary to promote Kepler's three laws, Newton's mechanics, Maxwell's electromagnetism, Yukawa's theory, Dirac's anti-particles, Heisenberg's uncertainty principle, and Einstein's theories. They can stand by themselves and need no promotion.

Why? Because these are repeatably verifiable science. Many technologies such as GPS are based on the understanding of these foundational fields of science. The very fact of promotion with fervor exposes the quasi-religious agenda of some evolutionists.

Philosopher of science and atheist Michael Ruse seems to agree. He reflects on the history of macroevolution. He wants to address a question. "Is evolution a secular religion?"[44] He acknowledges the quasi-religious aspects of evolution in Darwin's day. Even today, he writes,

> There is professional evolutionary biology: mathematical, experimental, not laden with value statements . . . Then, sometimes from the same person, you have evolution as secular religion, generally working from an explicitly materialist background . . .

He seems to suggest that big evolution is a "secular religion."

3.7.3 Intolerance to Doubt Darwin

Criticism of Darwinism and its many offshoots is usually met with robust responses from ardent believers. Take, for example, the book *Charles Darwin: Victorian Mythmaker* written by a veteran author A. N. Wilson. He started the writing project with the position of a believer of Darwinism. He wrote in the Prelude, ". . . it did not even cross my mind that I would

43. Sheffield and Bauer, "Darwin," 10:10, and Masci, "Darwin," para. 1–3.

44. Ruse, "Evolution," 1523–1524.

come to disbelieve in Darwin's theories."[45] He apparently changed his position after his research on Darwin's life.

Predictably, Wilson's biography was met with harsh criticisms. For example, Jerry Coyne, a professor emeritus in the Department of Ecology and Evolution at the University of Chicago, wrote a book review published by the Washington Post. He writes,

> If we exclude books written by creationists—a group that A. N. Wilson doesn't identify with—"Charles Darwin: Victorian Mythmaker" is by far the worst . . . Wilson's claim is grossly misleading and, frankly, ridiculous.[46]

It is interesting to note that Coyne has already flushed the "books written by creationists" down the toilet.

I don't know of any scientist who is so robustly defended, not even Einstein—ubiquitously recognized for his many experimentally proven ideas on how Nature operates at present. His theories are revolutionary. By contrast, Darwin's theory is a philosophical worldview of past history.

Another example is Thomas Nagel. The renowned philosopher, who is an atheist, wrote a book called *Mind and Cosmos* to enunciate his arguments for the proposition summarized by the subtitle: *Why the materialist Neo-Darwinian conception of Nature is almost certainly false.*

Nagel argues that Darwinism can't explain the emergence of life, consciousness, and cognition. If Darwinian models can't explain how life began as Nagel argues, his book will pronounce the demise of Darwinism. Nagel concludes his philosophical analysis as follows.

> In the present climate of a dominant scientific naturalism, heavily dependent on speculative Darwinian explanations of practically everything, and armed to the teeth against attacks from religion, I have thought it useful to speculate about possible alternatives. Above all, I would like to extend the boundaries of what is not regarded as unthinkable, in light of how little we really understand about the world. It would be an advance if the secular theoretical establishment, and the contemporary enlightened culture which it dominates, could wean itself of the materialism and Darwinism of the gaps—to adapt one of its own pejorative tags. I have tried to show that this approach is incapable of providing an adequate account, either constitutive or historical, of our universe.[47]

45. Wilson, *Charles Darwin*, prelude.
46. Coyne, "Attack," para. 1.
47. Nagel, *Mind*, 127.

As anticipated, Nagel's book stirred up a hornet's nest. Prominent evolutionists were quick to denounce his apostasy for betraying materialism. *National Post* evaluated the reception of *Mind and Cosmos* as "vicious."[48]

Among many others, a lucid description of the vicious responses to Nagel's book is found in an article by Andrew Ferguson, a senior editor at The Weekly Standard. Ferguson describes a leading atheist Daniel Dennett, who was frustrated and saddened by "this sort of retrograde gang, going back to old-fashioned armchair philosophy with relish and eagerness." Dennett complained, "It's sickening. And they lure in other people. And their work isn't worth anything—it's cute and it's clever and it's not worth a damn."[49] Alex Rosenberg, author of *The Atheist's Guide*, was quick to add, "And then there's some work that is neither cute nor clever, and it's by Tom Nagel."[50]

If Neo-Darwinism is almost certainly false as Nagel frankly proclaims, where else can materialists turn to for the intellectual foundation of their belief? You can see that there is a conflict of interest here when they comment on Nagel's book. No logical flaw in Nagel's arguments was brought up.

Here we see that most staunch believers of Darwinism who are supposedly rational can become emotional when intellectually honest arguments are made against Darwinism. They are also against Christian belief in the Creator of everything, labeling Christians as being irrational, emotional, dogmatic, and so on. But wait a minute. Here, we find that these atheists were emotional, dogmatic, and intolerant, when an established atheist philosopher openly denounced Darwinism.

Of course, neither the Christian belief, nor Darwinism, nor Nagel's alternatives are scientifically testable. Scientific approaches, like any other human endeavors, have their limits. Scientific methods are not meant to test beliefs about the historical *origin* of life and biodiversity. They are meant to test hypotheses by many experiments at the present time.

Therefore, ultimately, concerning the natural *history* of origin, it boils down to belief. Christians need to exercise faith in the almighty God who said that He created and made everything. Atheists need to have faith in the purely naturalistic philosophy and dark deep time. So it is germane to label Darwinism as a philosophical belief. Michael Ruse describes it as a "secular religion."

48. National Post Staff, "What," para. 3.

49. Ferguson, "Heretic," para. 11.

50. Ferguson, "Heretic," para. 14.

3.8 Is Science in a Crisis?

In March 1988, I graduated with a master's degree in science. I delayed my departure on a one-way ticket back to Singapore, as I had applied to the Public Service Commission (PSC) of Singapore Government to permit me to continue with the PhD program with scholarship.[51] I was waiting for the reply.

When my wife[52] and I were strolling in the Yoyogi Park, we found ourselves in the flurry of falling cherry blossom petals. It dawned on me that PSC still wouldn't approve my staying on to complete a PhD research program, as I had written an appeal letter to them. So when the final negative reply from PSC arrived, I was mentally prepared to accept the crisis in my adventure in science.

Reluctantly, I returned to Singapore and started to serve in the army for the balance of a year and a few months of my National Service duty. After the run-out-date as it is called in the army, I was posted by PSC to the Ministry of Defence and then to the DSO National Laboratories in 1989 as a defense science engineer. I was to work in the Software Lab then.

I proposed to do a project on recognizing handwritten numbers with rotation invariance and parallel processing. This project allowed me to be stationed at the Image Processing Lab set up by Professor Hock Lim. The Lab was housed in the Physics Department. The reason for this project was for the most part, I had obtained the consent of Professor Choy Heng Lai to do a part-time PhD program in theoretical physics under his supervision.

3.8.1 Self-Aware Machines?

The late 1980s saw the resurgence of neural network approach to artificial intelligence (AI). In the past, programmers had to code the rules manually. The neural network approach is totally different. Instead, we train a network of interconnecting nodes to learn from data how to perform a very specific task.

An example is machine translation. Using a computer language called LISP, AI researchers used to program the task of translation as complicated if-else rules. By contrast, a neural network learns how to do it by adjusting

51. Although my scholarship was from the Japanese Government, PSC had the final say.

52. I was married in the second year of my master's degree program. We met in a church during my senior year. The additional two years gave me the time needed to obtain the approval from her parents.

each weight of the connection from one node to the other node. A particular learning algorithm is known as back propagation. It is one of the many methods of making a machine learn from a large amount of data.

This neural network approach was something new in the Software Lab I was affiliated to at that time. I learned neural network as quickly as possible. My intention was to deliver a software prototype for a proof of concept ahead of schedule. Then I could focus on my research in theoretical physics.

By serendipity, I zoomed in on an artificial neural system called Neocognitron. It was invented by Fukushima in the late 1980s. His original model could recognize binary handwritten characters. I developed an adaptive algorithm that allows Neocognitron to recognize grayscale images as well.

Journal publication wasn't in my job description as a defense science engineer. But I was motivated to publish. I managed to publish a paper in 1993 under the pseudonym "Christopher Ting."[53] I also published another paper on the "Magnocellular Pathway for Rotation Invariant Neocognitron" in another journal.[54]

Having done an in-depth research on simulating physiological vision, I realized that computers were dumb. At the substrate level, computer consists of myriad on-off switches. These switches are based on the physics and chemistry of semiconductor. It is quite nonsensical to believe that Neocognitron has any consciousness when it is made to recognize patterns. I was busy programming Neocognitron in parallel C language, so much so that I was oblivious to the view of a giant in AI—Marvin Minsky.

John Horgan—a celebrated science writer—reported his interview with Minsky. Horgan writes,

> He expressed contempt for those who doubted whether computers could be conscious. "Consciousness is a trivial issue," he said. "I've solved it, and I don't understand why people don't listen." Consciousness is merely a type of short-term memory, a "low-grade system for keeping records." Computer programs such as LISP, which have features that allow their processing steps to be retraced, are "extremely conscious," more so than humans, with their pitifully shallow memory banks.[55]

I certainly would be a subject of Minsky's contempt. But isn't Minsky's definition of consciousness as a type of short-term memory too simplistic? In his definition, your mobile phone would have consciousness. But can

53. Ting and Chuang, "Adaptive," 285–299.
54. Ting, "Magnocellular," 43–54.
55. Horgan, *End*, 187.

soulless computers be conscious through the process of shuffling digital symbols? Does a non-living computer really have a mind? How could Minsky scientifically find self-awareness of a computer? Is there any scientific evidence for the material existence of mind?

3.8.2 Afterlife?

Francis Crick, after working on the double-helix DNA and further, switched gear to move in a totally different direction. He proposed what he called the Astonishing Hypothesis, which states that "a person's mental activities are entirely due to the behavior of nerve cells, glial cells, and the atoms, ions, and molecules that make them up and influence them."[56] This hypothesis exudes reductionism and materialism to its core.

But today, research papers are piling up, which suggest something contrary to Crick's proposition. Consciousness is not reducible to only the activities in the neural circuits of the brain. Quite the opposite, a rich literature of research papers has shown that mental activities can change the brain. It's mind over matter. For example, Miller et al. reported in 2014 that

> A thicker cortex associated with a high importance of religion or spirituality may confer resilience to the development of depressive illness in individuals at high familial risk for major depression, possibly by expanding a cortical reserve that counters to some extent the vulnerability that cortical thinning poses for developing familial depressive illness.[57]

In simpler terms, people who are prone to depression and who are religiously spiritual develop a thicker cortex that makes them less likely to be drawn into depression. "The importance of religion or spirituality appears to confer a neuroanatomical resilience in those who are otherwise predisposed to developing depressive illness."[58]

Science has falsified Crick's Astonishing Hypothesis.

After criticizing philosophers for having "such a poor record over the last two thousand years that they would do better to show a certain modesty rather than the lofty superiority that they usually display,"[59] Crick went on to criticize Christianity. He wrote,

56. Crick, *Astonishing*, 271.
57. Miller et al., "Neuroanatomical," 128–135.
58. Miller et al., "Neuroanatomical," 134.
59. Crick, *Astonishing*, 258.

> Not only do the beliefs of most popular religions contradict each other but, by scientific standards, they are based on evidence so flimsy that only an act of blind faith can make them acceptable. If the members of a church really believe in a life after death, why do they not conduct sound experiments to establish it? They may not succeed but at least they could try.[60]

This Nobelist was espousing his worldview of scientism. It is an exaggerated trust in the efficacy of the methods of natural science applied to all areas, including philosophy, the social science, and the humanities.

Crick was right, though, to remark that beliefs of different religions contradict each other. But this is a trivially self-evident statement—for otherwise we would have only one religion if the beliefs were consistent with each other. Crick was wrong, however, to assert that the belief in Jesus is an "act of blind faith."

Crick challenged Christians to produce the evidence of a life after death. In the Bible, we have *eyewitness* accounts of Jesus's bodily resurrection. Paul encountered Jesus on the road to Damascus. Additionally, in the transfiguration, the gospel writers also recorded that the transfigured Jesus was talking with Moses and Elijah. Moses died about fourteen centuries before the three eyewitnesses, Peter, James, and John were born. Elijah went up by a whirlwind to heaven eight centuries before their births.

Most importantly, Jesus affirmed the afterlife when He corrected the Sadducees. They, like Crick, Hawking, and many atheists today, didn't believe in life after death. Jesus said,

> But regarding the fact that the dead rise again, have you not read in the book of Moses, in the passage about the burning bush, how God spoke to him, saying, 'I AM THE GOD OF ABRAHAM, AND THE GOD OF ISAAC, AND THE GOD OF JACOB'? He is not the God of the dead, but of the living; you are greatly mistaken.[61]

By quoting Exodus 3:6 (and also 3:15), and by stating that God is God of the living, Jesus told the *fact* that Abraham, Isaac, and Jacob were living in the afterlife. Crick was "greatly mistaken."

Jesus's tomb is empty, and the address of His "afterlife" in heaven is "at the right hand of God." So Christians do have concrete evidence and documentation of life after death.

60. Crick, *Astonishing*, 258.

61. Mark 12:26–27 and Luke 20:37–38.

Crick's suggestion of doing experiments on life after death doesn't adhere to the ethical requirements when experiments involve human subjects. Crick has passed on and he would have found out whether there is an afterlife. Or, according to his belief, he would be dust with no consciousness. The end.

I wonder how would his fellow atheist Stephen Hawking respond if Crick hypothetically challenged him, saying, "Steve, if you really believe in radiation from a black hole, why do you not conduct sound experiments to establish it? You may not succeed but at least you could try."

If Hawking were modest, he would probably concede that the black hole radiation named after him is too weak. It's indistinguishable from noise. So, even in some cases of science such as the Hawking radiation, it is practically impossible for observations to be carried out. If it is almost impossible to conduct an observation in practice, but not in principle, should Crick then write off Hawking's theory of black hole radiation? Many theorists would likely protest.

The culture today treats ancient narratives in the Bible as scientifically guilty unless proven innocent. On the other hand, speculations on the historical origin of Nature dark deep time ago are treated to be scientifically innocent unless proven guilty. That said, neither life after death, nor the very process of life from non-life over dark deep time, can be ascertained by scientific observations. Both beliefs make the same claim of life from no life. Both beliefs are well beyond the methodology of science.

Therefore, nobody should be intimidated by scientists' speculative claims on things outside their areas of expertise. When scientists do so on matters crucial to Christianity such as afterlife, they speak not in the capacity of a scientist, but as any man in the street. Famous scientists are authorities in their respective fields. They are certainly intelligent. But their authorities in their narrow fields don't necessarily extend to matters outside their areas.

How is science today? Not very good. Some top scientists left their fields of expertise behind and did a poor part-time job in philosophy and theology with their scientism. They bring science into a midlife crisis.

4. What are the Prime Examples of Operational Science?

Theoretical physicists accept the need for mathematical beauty as an act of faith. There is no compelling reason for it, but it has proved a very profitable objective in the past. For example, the main reason why the theory of relativity is so universally accepted is its mathematical beauty.[1]

—Paul Dirac

. . . in the discussion of natural problems we ought not to begin at the authority of texts of Scripture, but at sensible experiments and necessary demonstrations.[2]

—Galileo Galilei

4.1 The Standard Model of Matter

In the last few hundred years, operational science has grown exponentially. From viruses to black holes, science has gone beyond our intuitive understanding of the things around us. Much of modern science deals with objects that can't be captured by our limited five senses anymore.

Since Newton's time, physics has become extremely broad in scope and far-reaching in its applications. We can see the division into theoretical physics, experimental physics, and computational physics. Nevertheless, physics is still about how physical things work, operate, and behave at present.

1. Dirac, "Methods," 22.
2. Ferguson, "Penalties," 358.

Experimental physics requires the ingenuity of physicists and engineers in scientific instrumentation. To illustrate this point, imagine that you want to carry out an experiment to investigate whether atoms can be frozen to a temperature of absolute zero Kelvin—about 273 °C below the freezing point of water.

What is temperature? It is the kinetic energy of atoms in a closed system. So you need to design and make a special container. You place the container on a platform with instruments that can neutralize any slight vibration from the floor. Otherwise, a tiny change in temperature could be due to the floor vibration and not to your input to the system. Also, how do you measure the temperature near absolute zero? Instrumentation of these devices requires a different kind of knowledge and skills.

One of the most expensive instruments in the world is the Large Hadron Collider (LHC). With a circumference of twenty seven kilometers, this giant machine looks like a big ring. The LHC is located 175 meters underground. To accelerate the sub-atomic particles ever closer to the speed of light, experimental physicists rely on the superconducting magnets installed at a facility of the European Organization for Nuclear Research (CERN).

The technologies for creating ultra-high vacuum, super-sensitive particle detectors, multiple cryogenic systems to cool the magnets, and many others over the 27-kilometer ring in this multi-national facility are perhaps second to none. High-precision engineering of the highest degree is the norm. Unlike Newton, who could conduct experiments on optics all by himself, scientists in the hundreds have to team up to run the experiments.

Suppose two bullets have to make a head-on collision. Even sharp shooters will find it extremely difficult to make it happen. Now imagine if the size of the bullet is shrunk by 100 trillion times. Even harder it is for two beams of protons to collide near the speed of light. Unprecedented is the level of high precision engineering over miles and miles of vacuum tube tunnel.

Most important of all, scientists need funding from sponsors of research projects. Often, government-linked agencies evaluate their proposals and provide funding to those with a track record of publication. More recently, scientists have also knocked on the doors of private companies. With very few exceptions, for-profit technology giants aren't keen to accept project proposals that seek to answer big questions. For example, what is the most fundamental structure of matter?

The LHC was built at an estimated cost of about five billion dollars, with an annual operating budget of nearly one billion dollars.[3] On top of

3. Knapp, "How," para. 4.

that, for carrying out the experiments, an additional 5.5 billion dollars per year is needed. Taxpayers pay for these experiments.

The huge investment paid off. Four years after LHC was turned on, CERN's CMS team announced in 2012:

> CMS observes an excess of events at a mass of approximately 125 GeV[4] with a statistical significance of five standard deviations (5 sigma) above background expectations. The probability of the background alone fluctuating up by this amount or more is about one in three million.[5]

Another team of CERN, the ATLAS group, reported in the same year:

> Both channels show a statistically significant excess at about the same place: a mass of around 126 GeV. A statistical combination of these channels and others puts the significance of the signal at 5 sigma, meaning that only one experiment in three million would see an apparent signal this strong in a universe without a Higgs.[6]

What is a Higgs? More properly, the question should be "Who is Higgs?" Well, Higgs is a British theoretical physicist. He published an important paper in 1964. His quantum field theory allowed him to make "the prediction of incomplete multiplets of scalar and vector bosons."[7] The so-called "scalar boson" is the missing particle named after Higgs.[8]

What do these two announcements mean? They mean that, finally, scientists at CERN have found the elusive Higgs particle. At the mass of 125 GeV, the Higgs particle is the heaviest among the fundamental particles of matter and energy. That's why LHC was built to produce enough energy to observe it. With these experimental achievements, the edifice of the standard model of particle physics, as shown in Figure 4.1, is now complete.

4. An electron volt, or eV, is the kinetic energy an electron initially at rest in vacuum will gain when an electrical potential difference of one volt is applied. Using Einstein's famous $E=mc^2$ formula, the mass m of a particle of one eV of energy E can be found. GeV is one billion electron volts.

5. CMS Experiment, "Observations," para. 2.

6. ATLAS Collaboration, "Latest," para. 4.

7. Higgs, "Broken," 508–509.

8. The symmetry breaking mechanism was also proposed by Englert and Brout ("Broken," 321–323), and Guralnik et al. ("Global," 585–587) in 1964 independently.

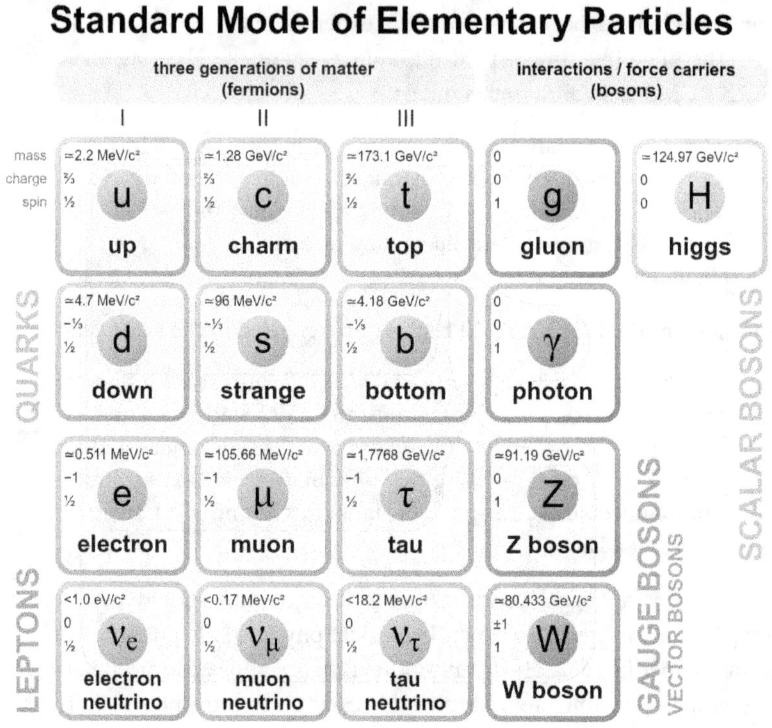

Figure 4.1. Standard model of fundamental particles: the twelve fermions and three interactions mediated by bosons. Credit: Cush. https://en.wikipedia.org/wiki/File:Standard_Model_of_Elementary_Particles.svg.

Next, what does sigma mean? The Greek letter sigma is the usual symbol in statistics to measure how far an observed value is from the average value. Let's assume that the observed values are randomly distributed in such a way that they plot a bell-shape curve. Then one unit of sigma is a statistic to indicate that the chance of the observed value being statistically no different from the average is 31.7 percent. Two units of sigma reduce that probability to 4.6 percent, i.e., about one out of twenty two chance that the observed value is a fluke.

The occurrence of a statistical fluke at 2 sigma is more likely than throwing two fair dice and get a double six. So 2 sigma is too loose a criterion. What about 5 sigma? The probability of seeing a statistical illusion is less than one out of three million. Put differently, it is as if you could break a six-digit combination lock by just one attempt. With 5 sigma, scientists at

CERN are 99.999943 percent sure that the Higgs particle is a reality. With 5 sigma, physicists regard the Higgs particle as a real particle.

The discovery of a "missing" particle requires deep engineering funded by deep pocket. By contrast, big bio evolution needs dark deep time. The "missing" links are still missing.

Although at an enormous cost, the elusive Higgs particle, thought to be "missing" previously, can be manufactured at present. Scientists have the technology to do so. CERN's discovery of the Higgs particle is a prime example of operational science. It is the result of a symbiotic collaboration between experimental and theoretical physicists.

Since J. J. Thomson's discovery of electron in 1897, it took particle physicists about one hundred and fifteen years to complete the full suite of empirical evidence for the standard model of fundamental particles. This model is based on quantum field theory, where localized quantum particles and spatially infinite quantum fields are two sides of the same coin.

We probably have heard of molecules, which are made of atoms. And we probably have learned that an atom is made of electrons and a nucleus. In turn, the nucleus is made of protons, which are positively charged, and neutrons, which have no electric charge.

Particle physicists realized that protons and neutrons are not the most fundamental particles. A proton is made of three quarks—two up quarks and one down quark. A neutron is made of two down quarks and one up quark. Other particles, such as Yukawa's pion, are made of quarks too. With quarks, we can elegantly classify all exotic particles produced by experiments.

Take a look at Figure 4.1. The fundamental particles display a beautiful structure of three families of matter. In each family, there is a pair of quarks, and a pair of leptons. These are the most fundamental constituents of matter known today.

Physicists are confident to tell us that the fabric of every atom and every particle in the universe consists of quarks and leptons. This is a shining example of reductionism at work. Can we trust the summary picture of particle physicists?

In principle, anyone can verify the theory and the corresponding experimental evidence at any place on Earth, and at any time. When we have repeatable testability, we have science—operational science.

On the other hand, the same can't be said for the theory of evolution. The chasm between dark deep time and present time is a showstopper. Biology is simply too complex, complicated, and full of exceptions. There are birds that can swim, and fish that can fly. Birds that can't fly, and fish that can't swim. Mammals that lay eggs, and reptiles that give birth. Plants that can move, and animals that don't move. Can the simplistic evolution theory explain these defying behaviors?

In Figure 4.1, we also see that there are three fundamental interactions in the standard model. The gluon is the strongest force among them. It binds the quarks together. Next is photon, which is the energy packet of the electromagnetic wave. Third in the list is the weak force; it explains the radioactive decay of atoms with the so-named W and Z bosons. The strong, electromagnetic, and the weak forces are the three fundamental forces that can be described by quantum fields.

The forth fundamental force of Nature, also the weakest, is gravity. It is captured by Einstein's general theory of relativity. So altogether, we have four fundamental interactions. Cosmologists assume that the physics of everything in the universe can be explained in terms of these four fundamental forces. But is there a fifth force? Nobody knows.

What then is the role of the Higgs particle, or should we say, the Higgs field? A field over all space in physics is like a huge lake. From time to time, ripples appear in the lake. The ripple is the analogy of a fundamental particle. The size of the ripple corresponds to its mass.

In the quantum field theory of quarks and leptons, all of the fundamental particles are modeled as having no mass. Just like the massless photon, any particle without mass must travel at the speed of light. But in reality, as shown in Figure 4.1, all the particles of matter are found to have mass. This is where the Higgs field comes in.

The photon does not interact with the Higgs field, and it has no ripple in the lake. Therefore photons are massless. By contrast, the W and Z vector bosons do interact with the Higgs field, and they cause big ripples in the Higgs lake.

It turns out that the neutrino that pairs with the electron is the lightest particle since its interaction with the Higgs field is negligible. So its mass is very close to zero. Many neutrinos come from the Sun. They travel near the speed of light. Solar neutrinos can pass through Earth's surface. A giant facility in Japan called Super-Kamiokande was specially built to detect neutrinos and study their behaviors.

At the significance level of 5 sigma, the experimental evidence for the validity of the standard model is obtainable at present time, and repeatably verifiable. Particle physics is a prime example of operational science.

4.2 Physics Envy

Big bio evolution can't match the experimental success of standard model of particle physics on all fronts. Big bio evolutionists seem to envy the glory of particle physics. Some of them like to use it to support their evolutionism.

An example is John Rennie. As the editor-in-chief, he published an article in the July 2002 issue of Scientific American. In it, Rennie seems to express his physics envy. He also has some misunderstandings of particle physics.

It is no mistake that Rennie is a staunch evolutionist. To him and many people, the reality of big evolution is established "beyond reasonable doubt."[9]

Rennie's arguments against "creationist nonsense" serve as a case study to discern the big difference between operational science and historical science. Below is a criticism of Rennie's arguments, not an attack of his personality.

Rennie uses the term "transformations" to refer to macroevolution, which is our big bio evolution. He claims that its "indirect evidence" is compelling. He writes,

> Although no one observed those transformations, the indirect evidence is clear, unambiguous and compelling.
>
> All sciences frequently rely on indirect evidence. Physicists cannot see sub-atomic particles directly, for instance, so they verify their existence by watching for telltale tracks that the particles leave in cloud chambers. The absence of direct observation does not make physicists' conclusions less certain.[10]

Rennie admits that "no one observed those transformations." He recognizes that macroevolution has not been observed. This is a fact on which we can agree. Nobody has seen a land animal becoming a whale.

Let's examine Rennie's argument. Drawing an analogy with particle physics, he argues that "the absence of direct observation" of sub-atomic particles makes "the indirect evidence" of macroevolution "clear, unambiguous and compelling."

But this loaded argument is inappropriate and misleading. It is loaded because Rennie assumes that sub-atomic particles can't be directly observed. This assumption is flatly wrong. Physicists have found that the proton looks like a black disk when its energy is ultra-high.[11]

9. Rennie, "15," 78.
10. Rennie, "15," 79.
11. Block and Halzen, "Experimental," 1–4.

4.2.1 Real Signals Always Have their Background

It's true that we can't see bacteria with the naked eye. But using a microscope, we can see them directly. We can magnify tiny objects, as I did in my master's degree program. I used a transmission electron microscope to take the images of complex proteins (see Figure 3.1). Similarly, we can't see distant galaxies. But using a telescope, we can. Astronomers often transform the electromagnetic signals from distant galaxies into images for us to see.

Microscopes and telescopes are instruments that detect and amplify the electromagnetic signals. Just like we can't see electricity, sub-atomic particles are invisible to us. Also, just like the electric current, their electric charges ionize the vapor in the cloud chamber. Thus, we can see trails of clouds as the results of their passages. The cloud chamber functions very much like a microscope.

There are something real that drive the trajectories in the cloud chamber. If sub-atomic particles aren't there, we won't see the mist-like trails, just like those parts of the cloud chamber where you don't see anything but the empty background. Yet at the same time, you do see some mist-like trails in other parts. The concomitant presence of the objects you are expecting to detect *and* their background is the hallmark of a real observation.

By contrast, we don't see big "transformations." As Rennie admits, they are not directly observable. We also don't see any background of intermediate lifeforms. The interpolating lifeforms that link dinosaurs to chickens are absent. Also, the fossils of "common ancestors" are unavailable. Without the background, how can we conclude that chickens came from dinosaurs?

4.2.2 Repeatable versus Non-Repeatable

Anyone can observe the trajectories of ionized vapors being formed in a cloud chamber over and over again. We can repeatedly see new ionized tracks. By contrast, the process of some primates evolving into some kind of bipedal animals can't be observed at present, let alone repeatedly. As Rennie says, "No one observed those transformations."

Origin is a historical event. For example, we may ask out of curiosity, "Where did the Japanese come from?" Historians have a few theories based on some ancient documents and artifacts. But historians can't conduct experiments to address the what-if questions.

Similarly, the origin of humans is a historical event. We can't repeat history with experiments. It had already happened, and it won't happen again. Just like history, origin isn't repeatable. Evolutionists seldom acknowledge this fact.

In any case, big evolutionists must believe that big "transformations" require many millions of years. This is the excuse for not seeing them today. Given long enough time, evolutionists say that lower lifeforms can become more complex. But "no one observed those transformations" because they take millions of years. It this science?

Since every "transformation" takes a very long time, the analogy with sub-atomic particles becomes invalid. These particles are observable *now*. We don't need to wait ten seconds for another new track to appear. We can repeatably see them in the cloud chamber as trails of ionizing charges.

Physics is a science of repeatable experiments. It has nothing to do with big evolution. Any association of big "transformations" with particle physics is misleading.

4.2.3 Cause-to-Effect versus Effect-to-Cause

In general, operational science involves experiments that begin with a cause. It is a set of conditions to produce the observable effects. For example, the cosmic ray of sub-atomic particle is the cause. The effect is the trail it generates in the cloud chamber. On the other hand, big evolution starts with the effects such as fossils, DNA, and so on. Using these effects, evolutionists then cook a cause, for example, the unknown common ancestor in the history of big evolution. Using the unobservable unknowns from dark deep time of the past to explain the effects is a definitive characteristic of historical science.

This effect-to-cause study is what historians are doing. With the artefacts and ancient documents as the effects, historians try to infer their plausible causes. Historians then write their versions of the history. Given the same "observation data" of World War II, the history written by Koreans and the one by Japanese are different. Likewise, given the same observations of Nature, the natural history of big evolution and the biblical history of creation are vastly different.

On the other hand, operational science is about cause to effect. When we specify the cause, we can see the effects unfold before us. We perform experiments to study how a cause brings about an effect. Here, cosmic rays from within our Milky Way Galaxy and from our Sun[12] are the *known* causes. The effects are the mist-like trails in the cloud chamber.

12. Friedlander, "Cosmic," para. 1.

4.2.4 Experiments versus No Experiment

Since the discovery of the Higgs boson, physicists have been studying its properties. This particle decays "almost instantaneously into other particles."[13] Even though the Higgs boson is invisible, particle physicists don't consider their experimental results as indirect evidence. They know *how* the Higgs particle decays into various particles. Recently, their experimental results have shown that the Higgs boson can decay into two muons.[14]

In other words, physicists can perform concrete experiments on the invisible sub-atomic particles. They can identify different particles. They can measure how strong is the signal. They can play with their giant instrument.

By contrast, evolutionists can't do experiments on the historical "transformations." Supposedly, they had already happened. Transformations can't be reversed for evolutionists to play with history.[15]

The four points above are not based on any religious belief. They show that a logical fallacy of false analogy has been committed in Rennie's article.

4.2.5 Implications of Big Bio Evolution

Let us now briefly consider a moral implication of big evolution. If humans did evolve from lower animals, and if humans were just animals, there is nothing morally wrong to kill other humans. When a hyena and a lion kill each other's babies, we don't condemn them for committing a hideous crime. We also observe lion killing the cubs fathered by a rival lion. Again, we find nothing wrong with that. Animals do what they have to do. We don't teach animals on the subject of ethics.

Some atheists such as William Provine believed in the absence of ethical absolute. He attributed his disbelief to evolutionary biology. In a debate between him and Phillip Johnson at Stanford University, April 30, 1994, he said,

> Now if I find myself aware after I'm dead, I'm going to be really surprised! But at least I'm going to go to hell, where I won't have all of those grinning preachers from Sunday morning listening.
>
> Let me summarize my views on what modern evolutionary biology tells us loud and clear—and these are basically Darwin's views. There are no gods, no purposes, and no goal-directed forces of any kind. There is no life after death. When I die, I

13. CERN, "CERN," para. 2.
14. CERN, "CERN," para. 1.
15. See Sarfati and Matthews, *Refuting* and Sarfati, *Greatest* for more arguments.

> am absolutely certain that I am going to be dead. That's the end of me. There is no ultimate foundation for ethics, no ultimate meaning in life, and no free will for humans, either.[16]

If mankind did evolve from lower animals, Provine's view of no ultimate foundation for ethics would be the rational and logical conclusion.

When Hitler rose to power, Nazis regarded Jews and Black people, and those with disabilities as inferior. Hitler had plans to decimate those residents of Germany who were regarded by the society as not fit for living. The history of mankind is marred by the Holocaust.[17] But this sort of killing is arguably consistent with evolutionists' worldview. If humans are animals, animals do what they have to do.

By decimating the weaker ones in the society, more resources can be channeled to the fitter ones. This policy would increase the chance of survival of humans against potential threats lurking somewhere within the planet—unknown unknowns. For example, gorillas somehow become more intelligent than humans suddenly. You never know.

Therefore, in the big evolution worldview, there is nothing morally corrupt to sterilize those people considered to be inferior. The scars of eugenics still remain today. More pertinently, under evolutionism, there is nothing wrong with "racial" discrimination either, because there is "no ultimate foundation of ethics."

It seems that Pekka-Eric Auvinen put this big evolution worldview into practice in our twenty first century. He was probably mentally ill and suicidal. Even so, he justified his killing this way. He wrote "I, as a natural selector, will eliminate all who I see unfit, disgraces of human race and failures of natural selection."[18]

Was he wrong? If there was no afterlife, this self-described "god-like atheist"[19] would get away with his crime scot-free.

On the other hand, by the biggest contrast possible, the Bible clearly states that humans were created in the image of God. But animals were not made in this way. Because God gave His divine breath of life (נִשְׁמַת חַיִּים *nishemat hayim*)[20] to humans only, we aren't animals. Though each of us was born under different circumstances in this fallen world, we are equal

16. Provine and Johnson, "Darwinism," para. 57–58.

17. For more details on the connection of Holocaust to Darwinism, see Bergman, *Hitler*.

18. CNN reporter, "Finland," para. 19.

19. CNN reporter, "Finland," para. 18.

20. Genesis 2:7.

on the basis of bearing the image of God. Pekka-Eric Auvinen had sinned against God and against the image-bearers he killed and wounded.

I hope this is an isolated case.

Also in Genesis, there is only one mankind, one race, as all ethnic groups originated from Noah's three sons. Whether Jews or Gentiles, whether Japanese or Germans, and regardless of our skin colors, we are all related through Noah, who was the descendant of Adam. We are distant relatives to each other. We are not the relatives of chimps. So in the biblical worldview, discrimination is wrong. There are absolute moral laws.

4.2.6 Misuse of Operational Science

Now, in answering "creationist nonsense," Rennie again refers to physics and writes:

> A central tenet of modern science is methodological naturalism—it seeks to explain the universe purely in terms of observed or testable natural mechanisms. Thus, physics describes the atomic nucleus with specific concepts governing matter and energy, and it tests those descriptions experimentally. Physicists introduce new particles, such as quarks, to flesh out their theories only when data show that the previous descriptions cannot adequately explain observed phenomena. The new particles do not have arbitrary properties, moreover—their definitions are tightly constrained, because the new particles must fit within the existing framework of physics.[21]

I have two remarks with regard to Rennie's understanding about particle physics.

One, particle physics *per se* doesn't seek to explain the universe. Particle physics takes the universe as given. It looks for natural laws, symmetries, and principles to *describe* mathematically its fundamental constituents and how they interact.

Physicists who venture into the cosmic beginning of the universe are big evolutionists. They apply particle physics to construct what they think is the historical origin. Their objective is to explain how the universe *came* into existence by the laws of Nature alone. This cosmological part is not particle physics. This part is historical science. It isn't the day-to-day operational science of particle physics.

21. Rennie, "15," 84–85.

Two, it is partially incorrect to say that physicists "introduce new particles, such as quarks, to flesh out their theories only when data show that the previous descriptions cannot adequately explain observed phenomena."

In the 1950s, experimental physicists discovered many exotic particles. They form a class of particles dubbed hadrons. The ordinary protons and neutrons are also hadrons. The esoteric hadrons are experimentally generated in particle accelerators. Although they are short-lived, we can still describe them by mass, electric charge, and various quantum properties. We can describe them with quantum physics and radioactive decay.

It's just that we have a variety of hadrons. Enrico Fermi, a rare genius in both theoretical and experimental physics, was quoted as saying, "If I could remember the names of these particles, I would have been a botanist."[22]

It is the diversity of hadrons that motivated theoretical physicists to develop classification schemes. None of them is as fundamental as quantum physics or general relativity. Gell-Mann and Zweig invented the quark model to account for all the hadrons in terms of six different "flavors." In this sense, it isn't quite accurate to say that "previous descriptions cannot adequately explain observed phenomena."

Having said that, Rennie is spot on when it comes to cosmology. The philosophy of big evolutionism is dominant in this historical science. Some cosmologists such as Stephen Hawking are trying to explain the origin of the universe with "nothing."

Big cosmic evolutionists need to introduce the unknown inflaton, which is a theoretical device to stretch the space faster than the speed of light. They also need to bring in the unknown dark matter and dark energy, "when data show that the previous descriptions cannot adequately explain observed phenomena." These unknown matter and energy are needed to patch up the model—"to flesh out their theories." These inflatons and dark things fit the "shadowy entities" coined by Rennie.[23] Without these obscure things, existing phenomena in the universe would deflate the Big Bang model.

Operational physics at present time is far more rigorous and quantitative than the imaginary drawings of evolutionists, who can't tell how many mutant apes were needed for humans to appear, by accident. There is nonetheless a similarity between the quark model and common ancestor model in that both are trying to explain diversity.

But this is where the similarity ends abruptly. The quark model has been experimentally proven. It gave specific predictions of new hadrons

22. Lederman and Teresi, *God*, 15.

23. Rennie, "15," 85.

yet to be discovered. Experimental results subsequently confirmed these predictions.

The chasm of stark differences is that ideas of macroevolution can't be tested with experiments; the processes of big evolution can't be observed at present. Evolutionists don't have the fossils of common ancestors. They can't predict "new" living things yet to be discovered. Dark deep time is always needed. This unbridgeable chasm disconnects big evolution from science.

4.2.7 Confession of Francis Crick

Francis Crick, an ardent atheist, and the co-discover of the double helix, wrote,

> Biologists must constantly keep in mind that what they see was not designed, but rather evolved. It might be thought, therefore, that evolutionary arguments would play a large part in guiding biological research, but this is far from the case. It is difficult enough to study what is happening now. To try to figure out exactly what happened in evolution is even more difficult. Thus evolutionary arguments can usefully be used as *hints* to suggest possible lines of research, but it is highly dangerous to trust them too much. It is all too easy to make mistaken inferences unless the process involved is already very well understood.[24]

For once, I can't agree more with an atheist that "to figure out exactly what happened in evolution is even more difficult." The irrelevance of "evolutionary arguments" is acknowledged. It is "far from the case" that they "would play a large part in guiding biological research." Indeed, during my experimental research in biophysics, "evolution" wasn't even in the vocabulary.

Isn't it quite bizarre that "useful" evolutionary arguments are "highly dangerous to trust them too much"?

It is also quite revealing that "biologists must constantly keep in mind that what they see was not designed, but rather evolved." Seeing design in biology yet mentally one must remind oneself that it isn't design. Do big evolutionists suffer from intellectual dissonance?

It goes without saying that the six-day creation of Genesis can't be verified with experiments. By the same token, the big "transformations" of Rennie can't be scientifically tested either. "No one observed those transformations." Seeing the effects today doesn't imply that the *true* historical cause

24. Crick, *What*, 138–139.

must be big evolution. One has to *believe* that the unobservable long history of big evolution did happen.

It is also very important to realize that observations need to be interpreted. For example, creationists may interpret the similarities in genomes as templates of common *design*. Big evolutionists may interpret them as evidence for common *ancestry*. Both interpretations can be wrong. But both can't be right at the same time. Which interpretation makes more sense?

The point is that the battle between creation and big evolution is not really Christianity versus science. It is the conflict of two worldviews. Since both worldviews refer to what had already happened in history, neither of them is testable by experiments. They are different interpretations of the artifacts. Both versions of history require faith—a belief in a version of history that nobody can see or experience.

If Bible didn't talk about God's creation at all, then there would be no contention. And, if we did see a lizard turning into a bird, as readily as tadpoles swimming in the water becoming a frog leaping on the ground, we would have an undeniable case for big evolution. Sadly for the evolutionists, this is not the reality. The missing links are still missing. We also see the oxymoron of "living fossils." These animals were supposed to have lost in the race for survival; they were once thought to be extinct by natural de-selection.

The Bible speaks with clarity about God's creation in Genesis and many consistent references to it throughout the Bible. Biblical creation used to be the default worldview of Jews and Christians, at least in the generation of Jesus and His disciples. The way Jesus referred to Genesis is historical. He said, ". . . from the blood of righteous Abel to the blood of Zechariah . . ."[25] Abel was a son of Adam who was murdered by his brother. Jesus interpreted Genesis literally.

Jesus must have been a creationist. This is hardly surprising, for He was the Creator in the beginning. The apostles appointed by Jesus were also creationists. Relatively recent big evolution is a challenge to the default creationist worldview—Jesus's worldview. If we believe in the miraculous resurrection of our body unto eternal life, can we turn away from His worldview—the miraculous creation without big evolution?

Creationists have always been criticized for using the supernatural to explain the mysterious. Indeed, applying god-of-the-gap arguments in operational science to explain lightning and thunder is really bad. But when it comes to historical matters, the situation is totally different. Creationists

25. Matthew 23:35, Luke 11:51.

attribute the origin of the universe to God. Big evolutionists assume dark deep time in their version of origin. It is a god of the gap of big evolution.

4.3 Phenomenology: A Case Study of Kepler

Within theoretical physics, there is a division. The first type is characterized by the works of theorists such as Dirac, Yukawa, and Higgs. The second type is called phenomenology.

The first type doesn't analyze experimental data at all. Mathematical analysis leading to fresh insight prevails. By contrast, phenomenology focuses on data analysis, such as the data churned out by CERN. In modern parlance, phenomenology of particle physics is data science plus the domain knowledge of high energy physics.

To illustrate the difference between theory and phenomenology, consider Newton's theory of gravity and Kepler's laws of planetary motions. Newton's theory is universal. It applies to all objects that have mass. Kepler's model, on the other hand, is phenomenological. It was invented primarily for describing the orbits of planets. In the Newtonian framework, Kepler's laws can be derived. Theory is the foundation of phenomenology in operational science.

Observation data of planets in our solar system are consistent with Kepler's three laws. Even if it takes Neptune about 165 years to complete one round, astronomers don't need to wait 165 years to confirm it. This is because Neptune's orbit can be reliably determined by Kepler's three laws. We can use the laws to predict its positions a year later.

Now, consider the observation data for the six planets in Table 4.1. These data were painstakingly collected by the Danish astronomer Tycho Brahe over a period of more than thirty years. The accuracy of the data was astounding for his time. In particular, the orbital periods were measured to the nearest sixtieths of a day.

Being an assistant to Brahe, Kepler eventually had access to the data. He used Brahe's data for a comprehensive analysis. In 1619, Kepler published his works in a book entitled *Harmonices Mundi* (*The Harmony of the Worlds*). It is a masterpiece.

For our intuitive understanding, I convert orbital period to years, and the average distance from the Sun to the astronomical unit (au). It is a system where Earth's average distance to the Sun is defined as one au. We take the average distance because planetary orbits are not circular but elliptical by the first law of Kepler. The average is calculated as the midpoint of the nearest distance and the furthest distance from the Sun. In absolute terms, one au is about 150 million kilometers.

Table 4.1. Tycho Brahe's observations of orbital period T and average distance R to the Sun.

	Orbital Period				Distance to Sun	
Planet	Days	1/60 day	Days in decimals	T (years)	Units	R (au)
Saturn	10759	12	10759.20	29.46	9510	9.510
Jupiter	4332	37	4332.62	11.86	5200	5.200
Mars	686	59	686.98	1.88	1524	1.524
Earth	365	15	365.25	1.00	1000	1.000
Venus	224	42	224.70	0.62	724	0.724
Mercury	87	58	87.97	0.24	388	0.388

At first glance, the measurement data in Table 4.1 don't seem to produce any obvious pattern. Nevertheless, it is clear that the orbital period is longer when the planet is further away from the Sun. We can also detect a positive correlation between the period T and the distance R. But this statement is qualitative at best, for it doesn't tell us how exactly a planet's period and distance are numerically related.

Kepler suggested that we should instead consider the cube of the distance from the Sun, R^3, and the square of the period, T^2. Let us plot these processed data points in Figure 4.2. It is evident that all the six planets line up nicely on a straight line. Included in Figure 4.2 is the zoomed-in view of the four planets nearest to the Sun. Again, we find that the same linear fit applies.

The straight line in Figure 4.2 is found by the best linear fit that goes through the origin and all six data points. That is, we regress R^3 on T^2 to obtain the slope of the straight line. It is found to be 0.9914—fairly close to the theoretical value of 1. The difference from the theoretical value is assessed to be statistically insignificant. Further statistical evaluation shows that the goodness of fit of the straight line to the data points is 99.9998 percent.

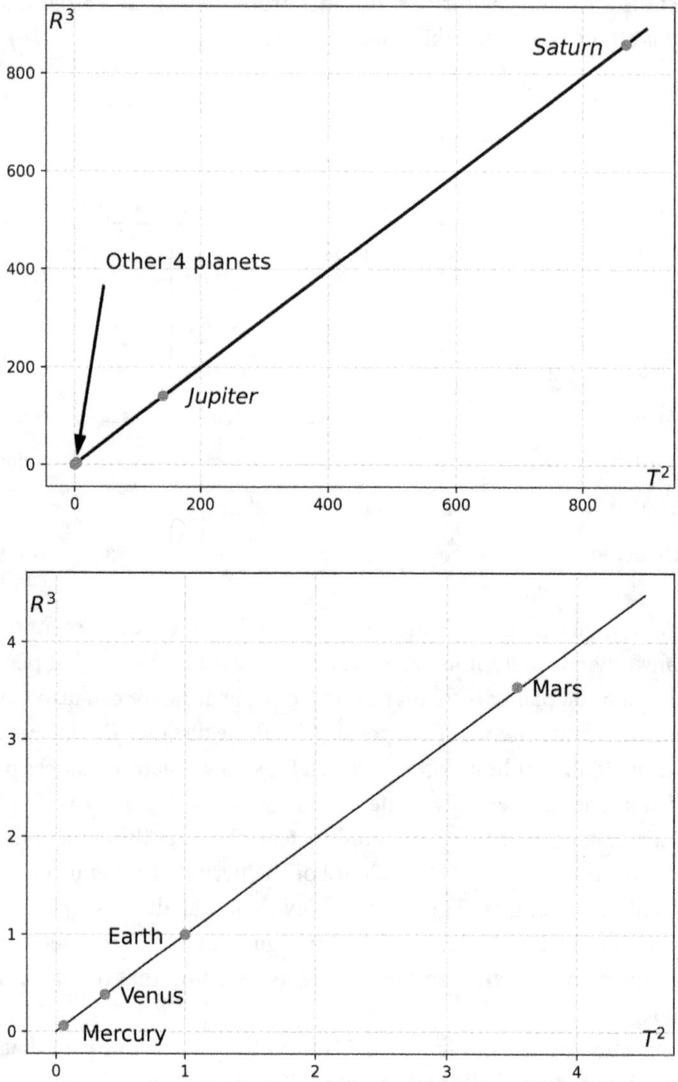

Figure 4.2. Kepler's third law.

The discoveries of Uranus and Neptune after Kepler had gone home to be with the Lord did not unsettle his three laws at all. It has been repeatedly verified that these two "new" planets follow Kepler's laws too. Kepler's three laws pass the test of time with flying colors.

But the same can't be said of big cosmic evolution beginning with a Big Bang. Fudge factors such as unknown inflaton, dark matter, and dark energy have to be introduced whenever the model doesn't fit new observations. Many speculative ideas have been proposed. Since neither dark matter nor dark energy interact with "normal" matter, the standard model of particle physics can't describe them. So no one has a clue of what these dark things actually are.

The reason for introducing dark matter is to rescue the model of galaxy evolution. The same can be said of dark energy, which was motivated by the unexpected discovery that the universe not only keeps on expanding but at an accelerating pace. The adjective "dark" is cosmologists' secret code for "mysterious, unknown, and unobservable."

Now, Kepler was a firm believer of God of the Bible. The following two quotes of Kepler, a protestant Christian, provide evidence for this fact.

> For we Christians cannot deny that the highest wisdom has presided over the instituting of the movements whereby the planet is made to run into its own region and is dispatched into its own spaces as if from the barriers; but Aristotle assigned this office to the movers themselves, as being eternal.[26]

> . . . I give thanks to Thee, O Lord Creator . . . in the works of Thy hands have I exulted. Behold now, I have completed the work of my profession, having employed as much power of mind as Thou didst give me . . . I have made manifest the glory of Thy works . . . if there is anything unworthy of Thy designs brought forth by me . . . breathe into me also that which Thou dost wish men to know, that I may make the correction: If I have been allured into rashness by the wonderful beauty of Thy works, or if I have loved my own glory among men, while I am advancing in the work destined for Thy glory, be gentle and merciful and pardon me; and finally deign graciously to effect that these demonstrations give way to Thy glory and the salvation of souls . . .[27]

Kepler's humbleness before God is refreshingly different from physicists such as Lawrence Krauss, Stephen Hawking, Steven Weinberg, and the rest who don't think God can possibly exist, let alone thank Him. Kepler was doing operational science. He wanted to show "the glory of Your works to man," thanking God for giving him all the abilities needed to do research.

In book V of *Harmonices Mundi*, having solved the final mystery of his lifework, Kepler was exuberant, and he wrote,

26. Kepler, *Epitome*, 51.

27. Kepler, *Epitome*, 240.

> . . . that most wonderful study began to shine . . . it is my pleasure to taunt mortal men with the candid acknowledgment that I am stealing the golden vessels of the Egyptians to build a tabernacle to my God from them . . . See, I cast the die, and I write the book. Whether it is to be read by the people of the present or of the future makes no difference: let it await its reader for a hundred years, if God Himself has stood ready for six thousand years[28] for one to study him.[29]

To Kepler, his scientific work was a platform to glorify God by solving a mystery of His creation.

4.4 Geocentric vis-à-vis Heliocentric Worldviews

There is a general misconception that the Bible upholds a geocentric view of the cosmos. But nowhere in the Bible does it say that the Earth is at the center of the cosmos. The Bible also doesn't say that the Sun moves around the earth.

It was the geocentric worldview of Aristotle (384–322 B.C.) that had a long-lasting influence on the natural worldview of people of antiquity. To Aristotle and many educated people, the geocentric worldview is intuitive. Nobody can feel the motion of the Earth except during a big earthquake. We also see the stars and moons moving across the sky. No wonder they used this geocentric worldview to interpret the Bible.

4.4.1 Danger of Scientism

Roman Catholic cardinals, popes, and educated people in the universities read into the Bible their worldview of Aristotle. Giorgio de Santillana, a philosopher and historian of science, writes,

> The natural philosophy of Aristotle, as well as the astronomy of Ptolemy adopted both by the universities and by the Church, taught that the Earth was at the center of things and that the heavens revolved around it in twenty-four hours, together with the Sun, Moon, and planets.[30]

28. In today's jargon, Kepler was a young earth creationist. Note that Kepler lived in the century before James Hutton and Charles Lyell came to the scene to popularize their worldview that the Earth was very old.

29. Kepler, *Harmony*, 391.

30. de Santillana, *Crime*, xxiv–xxv.

This belief in the natural philosophy serves as a sobering example of the evil of reading into the Bible ideas and claims never ever intended by the biblical authors.

Granted, there are passages that say that the Sun is rising such as Psalm 19 of David.

> The heavens are telling of the glory of God;
> And their expanse is declaring the work of His hands.
>
> . . .
>
> In them He has placed a tent for the sun,
> Which is as a bridegroom coming out of his chamber;
> It rejoices as a strong man to run his course.
> Its rising is from one end of the heavens,
> And its circuit to the other end of them;
> And there is nothing hidden from its heat.[31]

These verses suggest the rising of the Sun "from one end of the heavens to the other ends of them." The book of psalms belongs to the genre of poetry. This psalm says that the Sun has a circuit across the heavens. But don't all the meteorological stations use the same language of sunrise and sunset? The audience of this psalm and that of a weather forecasting station aren't students taking the lessons on Astronomy 101.

The church leaders of Galileo's day were doing eisegesis. They incorporated Aristotle's "scientific" worldview into the Bible. By instituting that no one should interpret the Bible except the church leaders, any worldview different from theirs was perceived as a challenge to their authorities. They misused David's psalm to support their geocentric belief. It is a clear case of misinterpretation.

In physics, you can set any frame of reference as the rest frame. This is part of Newton's first law. The geocentric worldview is one where the Earth is at rest, whereas in the heliocentric worldview, the Sun is considered to be fixed. In turn, with reference to the center of the Milky Way as the rest frame, the Sun is not fixed but moving. All these frames of reference have no connection with the central doctrines of Christianity. The belief in geocentric worldview is an additional doctrine, much like Mary's perpetual virginity.

Geocentric rest frame was useful for navigation with the constellations in the past. It is valuable also for building today's navigation satellite system. The global positioning system (GPS) on your mobile phone is enabled by having the Earth at the center with the satellites orbiting around it.

31. Psalm 19:1–6.

On the other hand, the heliocentric rest frame is convenient for describing the planetary orbits. Whether it is the satellites revolving around the Earth, or several planets going around the Sun, Kepler's three laws provide the best possible description of their trajectories. Nothing in this operational science of astronomy is in conflict with the Bible.

Nevertheless, the Inquisition of Galileo by the Roman Catholic Church cast a dark and long lasting shadow on Christianity. It perpetrates the polemic that "Christians are against science."

Some of the evolutionists have swung all the way to scientism. It is the ideology that science is the best or the only objective means by which society should determine normative and epistemological values. In short, scientism proclaims that science is the only way to truth. If you were to choose between science and Christianity, scientism would suggest that rational people should always choose science because not only is science more trustworthy but it also gets rid of blind faith.

But nothing is further from the truth. Like any other human endeavors, science has a limit, and scientists aren't almighty. It is a hubris to think that "mankind can surely conquer the heaven."[32] As an example, the nonlinear dynamics of the atmosphere is analytically unsolvable. Its mechanisms are so complex that gross simplifications have to be made in modeling the weather.

Humans can't stop famine, hurricane, tornado, hailstone, blizzard, heatwave, forest fire, drought, flood, tsunami, land slide, earthquake, volcano explosion, pandemic, and so on from happening. Despite all the breathtaking breakthroughs in technology, these disasters in our fallen world show how powerless humans are.

Science can't explain why there is something rather than absolutely nothing—no time, no space, no nothing. Science can't explain why there are laws of Nature for scientists to discover.

4.4.2 Galileo

Like Kepler, Galileo was a committed believer of the Scripture. Many Galileo's writings reflect his devotion to God of the Bible and to *directly demonstratable* science. A sample of his writings is furnished as follows.

32. This is a Chinese proverb.

1. I thank God from the bottom of my heart that he has pleased to make me the sole initial observer of so many astounding things, concealed for all these ages.[33]

2. The authority of the Bible was designed to persuade men of those articles and propositions which, surpassing all human reasoning, could not be made credible by science, or by any other means than through the very mouth of the Holy Spirit.

   Yet even in those propositions which are not matters of faith, this authority ought to be preferred over that of all human writings which are supported only by bare assertions or probable arguments, and not set forth in a demonstrative way. This I hold to be necessary and proper to the same extent that divine wisdom surpasses all human judgment and conjecture.[34]

3. But I do not feel obliged to believe that the same God who has endowed us with sense, reason, and intellect has intended us to forgo their use and by some other means to give us knowledge which we can attain by them. He would not require us to deny sense and reason in physical matters which are set before our eyes and minds by direct experience or necessary demonstrations.[35]

4. The intention of the Holy Ghost is to teach us how one goes to heaven, not how the heaven goes.[36]

5. The laws of nature are written by the hand of God in the language of mathematics.[37]

6. The prohibition of science would be contrary to the Bible, which in hundreds of places teaches us how the greatness and the glory of God shine forth marvelously in all His works, and is to be read above all in the open book of the heavens.[38]

7. . . . it seems to me very prudent of her to propose and of you to concede and to agree that the Holy Scripture can never lie or err, and that its declarations are absolutely and inviolably true. I should have added only that, though the Scripture cannot err, nevertheless some of its interpreters and expositors can sometimes err in various

33. Heilbron, *Galileo*, 153.
34. Drake, *Discoveries*, 183.
35. Drake, *Discoveries*, 184.
36. Drake, *Discoveries*, 186.
37. Lennox, *God's*, 23.
38. Hobbs, "Making," 443.

> ways. One of these would be very serious and very frequent, namely to want to limit oneself always to the literal meaning of the words; for there would thus emerge not only various contradictions but also serious heresies and blasphemies, and it would be necessary to attribute to God feet, hands, and eyes, as well as bodily and human feelings like anger, regret, hate, and sometimes even forgetfulness of things past and ignorance of future ones.[39]

Galileo is often said to be the first physicist. He was also a firm believer that God inspired all Scripture. He wrote, "For the Holy Scripture and nature both equally derive from the divine Word, the former as the dictation[40] of the Holy Spirit, the latter as the most obedient executrix of God's commands."[41]

As can be seen in quote 7 above, Galileo mentioned the grave consequence of not applying common sense in interpreting biblical passages. He went on to write, ". . . being obvious that two truths can never contradict each other, the task of wise interpreters is to strive to find the true meanings of scriptural passages agreeing with those physical conclusions of which we are already certain and sure from clear sensory experience or from necessary demonstrations."[42]

Notice that above all else, Galileo placed a strong emphasis on empirical observations and experiments, which he called demonstrations. His insistence on "a thousand experiments and a thousand necessary demonstrations"[43] flies in the face of big evolution with dark deep time of zero demonstration.

Notice also in quote 7 that Galileo appeared to oppose "literal" interpretation. If this quote were taken out of context, people who object six-day creation would probably use it as a bullet to argue for a non-literal interpretation of "day" in Genesis 1.

What Galileo meant by "literal" is that we shouldn't think that God literally has "feet, hands, and eyes, as well as bodily and human feelings like anger, regret, hate, and sometimes even forgetfulness of things past and ignorance of future ones."[44] Such heretic theology is contradictory to God being the Spirit. It is also blasphemous to His character of omnipotence, omnipresence, omniscience, and if I may add, omniperson (God as three

39. Finocchiaro, *Galileo*, 49–50.

40. It is not quite right to use the word "dictation." Holy inspiration is not dictation.

41. Galileo, "Letter," para. 5.

42. Galileo, "Letter," para. 6.

43. Galileo, "Letter," para. 8.

44. Galileo, "Letter," para. 4.

distinct persons but one being), immutability, impassibility, transcendence, holiness, infinite wisdom, and so on.

Galileo never mentioned anything from Genesis 1-11 at all. His opposition to the "literal meaning of the words" does not refer to the plain, straightforward, literal reading of Genesis as the history of the world, for Galileo was a creationist. As in quote 6, Galileo wrote about "the greatness and the glory of God" shining forth marvelously "in all His works."

4.5 Debunking the Myth of Science versus Christianity

It is a myth that Galileo's inquisition was about science versus Christianity. I was brought up with this teaching that Galileo's trial was about Christianity persecuting science. I didn't like Christianity when I was an atheist.

As I was exploring more, I found that it wasn't so simple. I shall debunk this myth of science versus Christianity. It is completely false. My assessment and conclusion concur with, for example, Lennox.[45] Keas went further to name some famous persons spreading the myth.[46]

The trial of Galileo had never been the case of science versus religion. As an example, Copernicus, living in a generation earlier than Galileo, was never persecuted for his proposal of heliocentric worldview. On the contrary, Copernicus's lifework was highly regarded by Cardinal Schonberg.[47] He dedicated his book to Pope Paul III.[48] Many myth spreaders overlook the honor Copernicus received from the religious leaders for his talent and his heliocentric model.

From what he observed with the telescope he made, Galileo concluded that the worldview of Copernicus was correct. The heliocentric worldview was contrary to the Aristotelian-Ptolemaic worldview widely adopted by university professors and educated people then. Among these believers of geocentric cosmos were some religious leaders who were supremely powerful politically. It was the era of Protestant Reformation when their authority was already under threat. The challenge to their interpretation of the Bible with the geocentric worldview was most unwelcome.[49]

Now, in a letter to Castelli, his friend and supporter, Galileo himself argued that it was the contest between his heliocentric worldview and the Ptolemaic world system.

45. Lennox, *God's*, 23–25.
46. Keas, *Unbelievable*, chapter 5.
47. Copernicus, *Revolutions*, xvii.
48. Copernicus, *Revolutions*, 3–6.
49. For a fuller account, see Statham, "Truth."

Specifically, in the biblical passage of Joshua 10, Galileo refuted the interpretation based on the Ptolemaic worldview. In this geocentric worldview, rather than the Sun stood still according to the Bible, it was something called the Primum Mobile that stopped.

The Primum Mobile is the outermost sphere in the universe known at that time. Its axial rotation every twenty four hours carries along all the other heavenly spheres and bodies in their daily motion from east to west around the motionless Earth at the center. Galileo argued from the physics of relative motions that it would be ridiculous for the Primum Mobile to stop in Joshua 10.

Galileo even gave a convincing argument that the Ptolemaic model was incompatible with the Scripture. In his lengthy letter to the Grand Duchess of Tuscany, Galileo gave an interpretation of the miracle in Joshua 10 as follows.

> The sun, then, being the font of light and the source of motion, when God willed that at Joshua's command the whole system of the world should rest and should remain for many hours in the same state, it sufficed to make the sun stand still. Upon its stopping all the other revolutions ceased; the earth, the moon, and the sun remained in the same arrangement as before, as did all the planets; nor in all that time did day decline towards night, for day was miraculously prolonged. And in this manner, by the stopping of the sun, without altering or in the least disturbing the other aspects and mutual positions of the stars, the day could be lengthened on earth—which agrees exquisitely with the literal sense of the sacred text.[50]

Galileo argued that *literally* as long as the Sun stopped, the Earth, along with the Moon and the rest of the celestial bodies, would remain in the same state for several hours. To him, the bottom line of Joshua 10 is this—it was a miracle, a divine intervention. The Primum Mobile in the Ptolemaic model is not needed at all. In fact, the Primum Mobile was anti-science, as it had impeded the progress of operational science.

It is also probably a myth that Galileo was always against the literal interpretation of the biblical narrative. Galileo cited Augustine in his letter:

> Now keeping always our respect for moderation in grave piety, we ought not to believe anything unadvisedly on a dubious point, lest in favor to our error we conceive a prejudice against something that truth hereafter may reveal to be not contrary

50. Drake, *Discoveries*, 213–214.

> in any way to the sacred books of either the Old or the New Testament.[51]

The key lesson Augustine gave us is that "we ought not to believe anything" based on "a dubious point."

As Galileo argued, using Joshua 10 and other passages in the Bible to condemn his heliocentric worldview was unjust. He exposed the prejudice of geocentric supporters. Galileo said the geocentric model was not only unscientific but also unbiblical. He asserted that it was his heliocentric worldview that was consistent with the *literal* interpretation of Joshua 10.

Today, few people have heard about the Primum Mobile, let alone believe in it. Could it be that the dark deep time of big evolution is the new Primum Mobile that has resurrected from the ashes of ignominy?

51. Drake, *Discoveries*, 175–176.

5. Is Big Bang the Historical Beginning of Time?

> Our fight is against some devilry that lies in the very *process* of things.[1]
>
> —Sir Herbert Butterfield

> Yet in another way, calculus is fundamentally naive, almost childish in its optimism. Experience teaches us that change can be sudden, discontinuous, and wrenching. Calculus draws its power by refusing to see that. It insists on a world without accidents, where one thing leads logically to another. Give me the initial conditions and the law of motion, and with calculus I can predict the future—or better yet, reconstruct the past.[2]
>
> —Steven Strogatz

5.1 What is Big Bang?

The origin of the universe is a mystery. Since antiquity, everyone wants to know how did the universe come about. Some cosmologists believe that the universe has no beginning and no end. But most people today believe that the universe *was* incredibly small, dense, and hot in the beginning. Since Big Bang is a past event, modern cosmology is a historical science.

Theorists must assume that all the laws of Nature have already been discovered and these laws alone are enough for writing an account of the origin. Afterall, cosmology is founded on Nature-only philosophy. It

1. Sewell, *Herbert*, 86.
2. Strogatz, *Calculus*, 1.

rejects any possibility of biblical creation. Big evolutionists think that their historical science can make people understand the history of the universe correctly. Some of them may say, "We have reliable observations. God has no role to play in the historical science of the origin." It is tantamount to "God didn't create."

5.1.1 First Secret

Many people have heard of Big Bang. The standard Big Bang cosmology can explain several modern observations when we *interpret* them in the framework of big evolution. Most scientists affirm the Big Bang cosmology because if offers the best explanation for what astronomers can observe today. But few people and scientists in other fields know the dark secrets behind Big Bang.

The first secret is Friedmann's equations. When the metric of space is assumed to be dependent only on a scale factor, Friedmann's equations are the special cases of Einstein's field equations of general relativity. If the universe is a sphere, the scale factor is some sort of its radius. In other words, the scale factor serves as a proxy for the size of the universe. Solving Friedmann's equations, we obtain the scale factor as a *deterministic* function of time.

Friedmann showed that the size of the universe increases with time. He predicted that the size of the universe will be larger than it is today. It means that the space of the universe is always expanding. More accurately, the intergalactic distance is getting longer and longer every day. Therefore, majority of the galaxies we observe today won't be observable anymore because they will become too far away for us to see with our best telescopes. This part about the future is perfectly okay. But whether this scenario predicted by Friedmann's theory will surely occur remains to be seen. Surely it won't happen in our lifetime.

Here comes the dark part of the first secret. Cosmologists argue that since the universe will become bigger, it must be smaller in the past. Like Newton's equation of motion, the Friedmann equations are deterministic. The deterministic property allows us to rewind time, pretty much in the same way when we play a video backward. This is extrapolation back in time. Extrapolation is unobservable.

If we reverse the arrow of time, the scale factor would become smaller and smaller, until it was infinitesimally close to zero. This part is a secret assumption. Cosmologists assume that Friedmann's solution always applied in the past. Also, the postulate is that the universe must have begun from an unimaginably tiny space. It was previously dubbed the "cosmic egg." Today, it is called "quantum fluctuation" or "nothing."

But could it be that Friedmann's calculus only worked until a time in the past when the scale factor wouldn't become smaller anymore? In the words of Strogatz, "calculus is fundamentally naive, almost childish in its optimism. Experience teaches us that change can be sudden, discontinuous, and wrenching. Calculus draws its power by refusing to see that. It insists on a world without accidents."[3] Since we still don't know a lot about our universe in the past, could accidents happen? Can unknown unknowns pop out to ruin the Big Bang model?

5.1.2 Second Secret

The Big Bang cosmology continues to capture the imagination of many people in the street. There appears to be a mysterious aura around it. Overwhelmed by the grandeur, many people believe and trust what cosmologists say about the universe. Some can even recite the mantra that the entire universe expanded from a dense, hot space no larger than a proton 13.8 billion years ago.[4]

The second secret is the appropriation of particle physics. This branch of physics is operational science. Particle physicists experiment with the most elementary constituents of matter and study their interactions. They can measure the mass of an electron. They can repeat the experiments.

Big evolutionists suppose that in the beginning, there were only these sub-atomic particles in the form of plasma. Lightening is an example of what a plasma is. So, in the beginning, big evolutionists think that the universe *was* incredibly hot. But they have never measured its temperature. They will tell you it is impossible to measure. That is, the temperature in the beginning *was* unobservable and unknown.

With the experimental confirmation of the Higgs particle, the standard model of particle physics is pretty much complete. This operational science shows us that at the quantum level, there are only three forces: the strong, the weak, and the electromagnetic. The so-called Yang-Mills gauge theory is the basic framework for these three. That's how Nature operates at the sub-atomic level. However, the Yang-Mills gauge theory is incapable of accommodating the fourth force of Nature—gravity.

Since the 1970s, theoretical physicists have moved ahead of the standard model. They conceived a grand symmetry where the three quantum

3. Strogatz, *Calculus*, 1.
4. Depending on the measurement methods, the "age" of the universe can vary.

forces are unified as one. However, the models of grand unified theory (GUT) are not experimentally feasible on Earth. The energy required to test these models is way too high. It is far beyond what the Large Hadron Collider can produce. Despite the lack of experimental evidence, cosmologists went ahead to appropriate the energy level above which these three forces are unified—in theory. They assume that at the split-second after the universe suddenly exploded, the energy *was* so high that the three forces of Nature *were* unified.

Here comes the dark side of the second secret. Big Bang cosmologists assume that one of the untested and currently untestable grand unified models is the true model. The ambition of Big Bang cosmology is to describe how the universe began to exist. Cosmologists and philosophers want to answer a big question: Why there is something rather than nothing?

Big Bang cosmology is about the natural history from the beginning of time when there *was* nothing to the present time. Since there is only one universe, the cosmic history must be unique, and the GUT model chosen must be unique as well.

But if Nature defies GUT theorists' expectation by refusing to unify, which isn't entirely impossible, the storyline of Big Bang must be rewritten. For the time being, cosmologists simply ignore this unresolved problem. They hope that at least one of a few GUT models would be good for the job. But since nobody can observe the GUT epoch, speculation about the early universe is probably no different from Darwin's warm little pond for the origin of life.

The cosmology about the origin is fundamentally different from operational science. Day-to-day science is built upon the solid foundation of experimental controllability, verifiability, and repeatability. None of these hallmarks that characterize a scientific field such as particle physics is found in the Big Bang model. By definition, history can't be controlled by keeping one variable fixed and the other varying. Neither can natural history be verified, as it is always billions of years of deep time.

5.1.3 Einstein's Gravity

What about gravity? Is it natural history? No, gravity is not history. You can test the inverse square law of gravity proposed by Newton in the lab today. From Newton's law of gravity, you can derive Kepler's laws, which are proven to be true, over and over again. This is operational science.

What is the role of gravity in the Big Bang model? To answer this question, we need to dive deep into the history of Big Bang. It all started with

Einstein's theory of general relativity. It is a generalized version of Newton's inverse square law of gravity. That is, if Einstein's general theory of relativity is simplified, you can recover Newton's model of gravity.

Einstein's field equations of general relativity describe the interplay between the geometry of spacetime and the motion of matter. Specifically, the matter makes the spacetime curved. In turn, the curved geometry determines the motion of matter. The key idea of Einstein field equations is that the feedback on each other is equal. The interaction between spacetime geometry and energy-matter distribution gives rise to a highly nonlinear structure.

Amazingly, up to a proportional constant, Einstein's field equations say that spacetime geometry is equal to the energy-matter distribution. These equations describe geometry and distribution as two sides of the same coin.

To make the universe static, Einstein added an additional term Λg. Here, g is a metric, namely, a mathematical construct of measurement for a particular geometric object. And Λ ("Lambda") is called the cosmological constant. The field equations become complicated.

In plain English, Λg is an anti-gravity device to counteract gravity. Like two magnets of opposite poles, gravity pulls two objects toward each other. By subtracting away Λg, the matter-energy level becomes lower, which means that the geometry becomes less curved, so less gravity.

Einstein believed that the universe was unchanging. His belief motivated him to introduce Λg to push away the pulling gravity of mass so that the universe will not collapse on itself.

5.1.4 Friedmann's Tricks

But could Einstein's cosmology be wrong? Gamow, who was once Friedmann's student, had a firsthand account of Friedmann's foray into cosmology. He wrote in his autobiography:

> Studying Einstein's publications on that subject from a purely mathematical point of view, Friedmann noticed that Einstein had made a mistake in his alleged proof that the universe must necessarily be stable and unchangeable in time . . . in the course of his proof, Einstein had divided both sides of one of his intermediate equations by a complicated expression which, in certain circumstances, could become zero . . . when this expression becomes equal to zero, Einstein's proof does not hold, and Friedmann realized that this opened an entire new world

> of time-dependent universes: expanding, collapsing, and pulsating ones.
>
> Friedmann wrote about his findings to Einstein . . . Einstein wrote Friedmann a short and somewhat grumpy letter, agreeing with his argument. Friedmann published his paper in 1922 in the German magazine *Zeitschrift für Physik*, thus opening a new era in cosmology.[5]

Gamow was right. Friedmann, who lived in the day of Soviet Union obscure to the Western world, was the earliest to suggest that the universe could expand. In his 1922 paper, Friedmann stated that the universe could be as old as ten billion years from time zero.

With a few clever tricks of math, Friedmann first simplified and reduced Einstein's ten field equations to only two differential equations. One of the tricks is to simplify the geometry of three-dimensional space to just one variable. This variable is called the scale factor. It is the most important variable in Friedmann's cosmology.

In other words, Friedmann assumed that space is isotropic. Direction is irrelevant. It means that the universe is a sphere like a basketball. The overall size of the universe is determined by the scale factor as if it is the radius of a sphere. This scale factor is assumed to be a function of time only.

Thus, we see that the Big Bang model requires the assumption of isotropic cosmic evolution. Namely, the universe is agnostic to any direction in the sky map. The universe looks the same in every direction far away.

Another trick of math to make things simple is to assume that the distribution of energy and matter is homogeneous. At the large scale, density variation of matter and energy is treated as background noise. Thus, we see that the Big Bang model requires the assumption of homogeneity at any given time as well.

Few scientists believe that isotropy and homogeneity are scientifically testable conditions when the universe was supposedly a three-minute-old baby. Baby universe is history. Nobody can go back to dark deep time to find out. However, without these simplifying assumptions, Einstein's field equations are almost unsolvable. This is because the ten equations of Einstein are highly nonlinear. Almost every variable is entangled with every other variable.

Given these assumptions, Friedmann, as an applied mathematician, reduced Einstein's field equations to just two ordinary differential equations. These two equations are deterministic in that you can compute or predict the size of the universe any time in the future and the past from today.

5. Gamow, "World," 44–45.

What does deterministic really mean? As an illustration and an analogy, consider the Moon. The regularity of Moon's orbit around the Earth allows you to predict its next position, say ten days later. So the orbit is already determined. The Moon has no choice but to be there as predicted.

It turns out that in Friedmann's framework, the simplest solution is a monotonically increasing function of time. It means that the universe will surely be bigger in the future. This is an unescapable destiny. The universe in Friedmann's model can't behave otherwise. It is fated for the universe to get bigger every day. Conversely, it also means that the universe *must* be smaller in the past. There would be a time in the past when the universe was as small as a pea. Does this extrapolation back to dark deep time make sense?

Calculus is fundamentally naive, almost childish in its optimism.

5.1.5 Einstein's Mistakes

Friedmann's 1922 solution simply says, "Dr. Einstein, the cosmological constant Λ is redundant." "Without it, the universe will keep on expanding anyway." "Don't worry. The universe won't collapse."

Interestingly, Einstein at first thought that Friedmann's solution was wrong. After reading Friedmann's paper submitted to the *Zeitschrift für Physik*, Einstein wrote to the editor:

> The results obtained in the cited paper regarding a nonstationary universe seemed suspect to me. In fact, it turns out that the solution given does not agree with the field equations . . . The significance of this paper therefore lies precisely in that it proves this constancy.[6]

Later, Einstein stood corrected. He wrote to the journal editor a year later:

> In my previous note I criticised [Friedmann's work On the curvature of Space]. However, my criticism, as I became convinced by Friedmann's letter communicated to me by Mr Krutkov, was based on an error in my calculations. I consider that Mr Friedmann's results are correct and shed new light.[7]

Genius makes mistake too. Einstein was biased by his worldview of a static eternal universe—no beginning and no end. Who can say the reviewers of historical science are one hundred percent objective?

6. Einstein, "Comment," 271–272.
7. O'Connor and Robertson, "Aleksandr," para. 23.

5.1.6 Implications of Big Bang

The greatest implication of Friedmann's solution is that the universe had a beginning. This is a ground-shaking result. Many philosophers and cosmologists believed that the universe was eternal, with no beginning and no end.

Among the people who thought this way was Einstein. His worldview biased his reading of Friedmann's 1922 result. To him, a non-static cosmos couldn't be right. So he described Friedmann's paper as "suspicious" and jumped to the wrong conclusion that Friedmann's solution didn't satisfy his field equations of general relativity. If Einstein wasn't blinded by his philosophical worldview of a static universe, he wouldn't make this mistake. Nonetheless, Einstein was courageous to admit his mistake.

As mentioned earlier, according to Friedmann's solution, the universe will expand with the flow of time. So if you could reverse the arrow of time, which is impossible in reality, it must contract until the scale factor is almost zero. The universe would have been just a point (called singularity). In a footnote, Friedmann wrote,

> The time since the creation of the world is the time that has flowed from that instant when the space was one point until the present state; this time may also be infinite.[8]

By saying that "this time may also be infinite," it seems that Friedmann wasn't sure whether the universe existed eternally in the past. On the other hand, it could also be that his worldview was still under the influence of Aristotle. This great philosopher of antiquity believed that the cosmos is eternal.

Nevertheless, in conclusion, Friedmann wrote,

> . . . the "cosmological" quantity remains undetermined in our formulae, since it is an extra constant in the problem . . . If we set $\Lambda = 0$ . . . then the world period becomes of the order of 10 billion years. But these figures can surely only serve as an illustration for our calculations.[9]

Serving only as an illustration, Friedmann half-heartedly speculated that the universe was about ten billion years old "since the creation." To many people, including Einstein, this worldview was contrary to the static and eternal universe many people believed at that time.

It is unknown whether Friedmann was a theist. Most probably not. Like Einstein, Friedmann nonetheless was a Jew by ethnicity. Limited as it might be, they probably had some exposure to the Old Testament. Having said that,

8. Friedmann, "Curvature," footnote 11.
9. Friedmann, "Curvature," footnote 11.

Einstein was definitely not a believer of God of the Bible. He didn't think that the miracles such as parting of the Red Sea and so on could have happened. In other words, Einstein didn't believe in miracles. He believed only in the laws of Nature for him to indulge in thought experiments.

So whenever Einstein used the word "God" in his speech and writing, it was not God of the Bible who can suspend the laws of Nature in a pinpoint fashion. For Einstein, it was a god reasoned to exist by a philosopher named Baruch Spinoza, who was a Jew too. Spinoza's god was so different from God of the Old Testament that he was "issued the harshest writ of herem, ban or excommunication, ever pronounced by the Sephardic community of Amsterdam; it was never rescinded."[10]

At any rate, Friedmann wasn't a religious Jew. For lack of a better word, he coined "creation" as an analogy to describe an implication of his solution. In the context of his writing, in no way did Friedmann mean the biblical creation. Nor did he mean the biblical beginning. It is unwise for well-meaning Christian apologists to use it.

5.1.7 Consequences of Embracing Big Bang

As if to appease the infamous Galileo affair, the Roman Catholic Church went an extra mile to support Big Bang and also big bio evolution. Another possible reason is that Georges Lemaître, one of the pioneers of Big Bang, was a Catholic priest. Interestingly, Lemaître presented to Einstein his solution of Einstein's field equations. The worldview of Einstein, however, was still a static universe. Lemaître received an abrupt comment from him. "Your calculations are correct, but your physics is abominable."[11]

Anyway, Lemaître succeeded in solving Einstein's field equations. He obtained essentially the same result of Friedmann. The physics abominable to Einstein is that Friedmann's solution leads to an expanding universe.[12]

Some well-meaning Christian apologists are quick to embrace the Big Bang cosmology. This historical science seems to offer "a beginning" that appears to fit "the beginning" of Genesis 1:1. "See, the Big Bang's beginning is consistent with the Bible. Even atheists admit there's a beginning." On the other hand, the possibility of a beginning might bring discomfort to some atheists.

But a careful analysis finds that the biblical beginning and Friedmann's theoretical beginning are totally different.

10. Nadler, "Baruch," para. 2.
11. Deprit, "Monsignor," 370.
12. Lemaître, "Homogeneous," 483–490.

Anyone who takes Big Bang's beginning literally and identify it with "the beginning" of Genesis 1:1 falls into a fallacy. This identification is theologically flawed. It implies that God can't be omnipotent and can't be sovereign. It means that God can't create.

Let me explain. The historical beginning of time is a mystery, both scientifically and theologically. As a matter of fact, we don't even know what time really *is*. We can accurately measure time with atomic clocks, but no one knows how time began. Time is the biggest mystery.

Earlier, I mentioned that Friedmann's solution is deterministic. When we rewind the clock, the scale factor of the universe must religiously follow Friedmann's solution. It has no other choice. The word "deterministic" means the lack of free will. It also implies that everything is already determined and no one can change the destiny.

If Christians were to embrace the secular Big Bang cosmology, then they would also have to accept the implication that God had no other choice. It means that God must begin His creation at the time determined by Friedmann's solution. God would be bound by a law of Nature before the creation. It means God didn't create the laws of Nature. It means that God is neither sovereign nor omnipotent.

Also, it would mean that God had to start His creation from a tiny universe no larger than a proton. But Genesis suggests that in the beginning the space wasn't a proton. At least the space was big enough to contain the Earth covered with water. Big Bang's beginning is inconsistent with the biblical account.

Moreover, a proton-size space is something. But God of the Bible created *ex nihilo*.[13] The transcendent God of the Bible created the world out of nothing—from no pre-existing matter, space, time, and laws of Nature. Nothing.

Can man's solution of some equations dictate a sovereign God's *ex nihilo* creation? If God was bound by the laws of Nature and He had no choice but to make space smaller than a proton in the beginning, then it wouldn't be the transcendent God of the Bible.

5.2 Redshift and Hubble's Interpretation

It is not an overstatement that the astronomer Edwin Hubble revolutionized mankind's worldview of the universe. Before his monumental work, people thought that the Milky Way was the universe. Hubble used one of the most powerful telescopes in the world then. He discovered nebulae beyond the Milky Way. The nebulae he observed are what we call galaxies today.

13. Worthington, "*Creatio*," 49–59.

5.2.1 Hubble's Statistical Law

Among the most important findings of Hubble is a "statistical law." It states that the farther apart galaxies are from us, the faster they seem to move away. This positive correlation is known as the Hubble law, which is somewhat similar to Ohm's law of electricity. Still, there is a crucial difference. Ohm's law is operational science. It can be explained by some underlying physics that involves Maxwell's electromagnetism.[14] On the other hand, Hubble's law is data science, as it is purely observational.

To illustrate Hubble's statistical law, let's consider the linear relation between the distance and the velocity of nebulae. It is the upward-sloping solid line in Figure 5.1. Each data point is plotted as either a circle or a cross. The distance is in the unit of mega parsec, which is about thirty one times billion of billion kilometers (km). The data are taken from Table 1 of Hubble's short paper he published in 1929.[15]

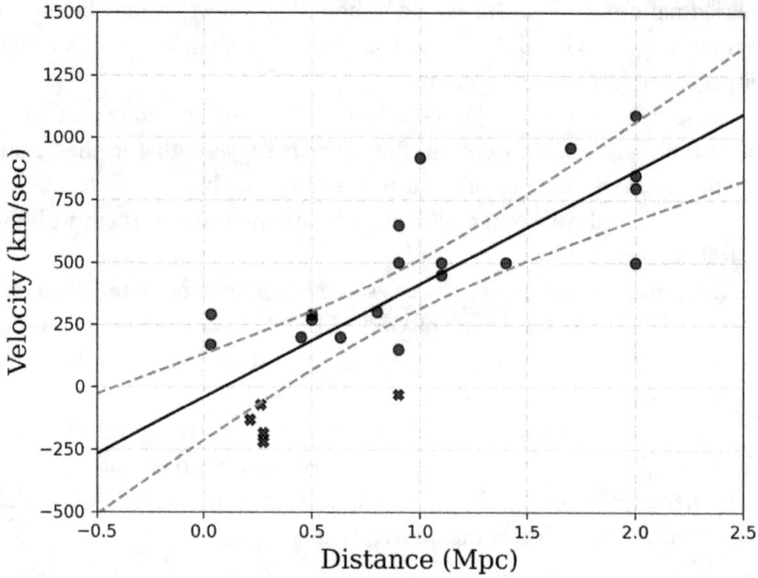

Figure 5.1. The data of distance and velocity are taken from Table 1 of Hubble's six-page revolutionary paper published in 1929. The solid line is the best linear fit for all the twenty four data points. The dashed curves demarcate the boundaries of the band. They capture the fitted value of velocity for any given distance with ninety five percent level of confidence.

14. Jang and Masmoudi, "Derivation," 3649–3669.
15. Hubble, "Relation," 169.

Now, the slope of the best fit line in Figure 5.1 is estimated to yield a value of 454.16. This gradient of the slope is dubbed the Hubble constant. The short form for the Hubble constant is H_0. The subscript "o" means the Hubble constant is the one measured today.

In contrast to Kepler's third law described in chapter 4, the adjusted goodness of fit is only 60.6 percent. As can be seen in Figure 5.1, there are a number of data points falling outside the band demarcated by two dashed curves. In particular, all the six galaxies indicated by crosses at the left bottom of the plot don't follow the Hubble statistical law.

The reciprocal of H_0 gives the age of the universe in the context of Big Bang cosmology. Hubble's estimate of 454.16 indicates that the age of the universe is about 2.2 billion years. This age is about six times smaller than the modern estimate of about 13.8 billion years, which corresponds to the Hubble constant of 67.74.[16]

Now, in the revolutionary 1929 paper, Hubble relied on the redshift. It is a phenomenon observed for the spectrum of light from a star or from a distant galaxy. The spectrum of light contains dark gaps or lines against the background of continuous wavelength. The dark lines that occur at the particular wavelengths allow us to identify the atomic elements in a star such as our Sun.

Is there any difference between the so-called absorption spectra from the nebulae and those from the Sun? Yes, there is a difference. The lines in the nebular spectra, in general, are not in their normal positions defined by those in Sun's spectrum. They are displaced from their normal wavelengths. The wavelengths corresponding to those dark lines are somewhat longer or shorter than those that you normally observe from the Sun.

These shifts in wavelengths were known to astronomers before Hubble published his 1929 paper. We can calculate the spectral shift by dividing the shifted amount with its normal wavelength. The resulting ratio varies from one galaxy to another. The shift toward the longer wavelength is called the redshift. Opposite to the redshift is the blueshift, where the displacement is toward the shorter wavelength relative to the normal absorption line of our Sun.

Without any interpretation, redshift and blueshift are what they are. Redshift is the lengthening of light's wavelength over the vast distance from the emitter to us. For a given galaxy, you can observe the same value of redshift repeatedly. This part is operational science.

The data points indicated by circles in Figure 5.1 come from Hubble's sample of red-shifted galaxies. Astronomers have also discovered galaxies that are blue-shifted. Indicated by crosses, the six data points in Figure 5.1 correspond to these galaxies.

16. Ade et al., "Planck," abstract.

Based on his observations of redshift and his novel way to measure the distance of nebulae from Earth, Hubble interpreted the observation of redshifts as the consequence of galaxies moving away from us. He then concluded that the universe expands *uniformly*. The implication is that the universe should be smaller in the past. And the universe is not eternal; it must have a beginning.

Hubble's monumental paper was instrumental in convincing many people to accept the worldview that the universe is changing. It only began to exist a finite time ago. But this worldview is predicated on his interpretation. Namely, redshift is the result of light emitters *moving away* from Earth. In this interpretation, it means that space itself is expanding. More accurately, it is only the space not occupied by any galaxy that is expanding.

5.2.2 Hubble's Choices

Under the cosmic Copernican principle, the concept of center of the universe is put to death. Our planet certainly is not the center of the solar system. Nevertheless, it isn't impossible for Earth to settle in a special location.

On this unwelcome prospect of Earth being special, Hubble told his audience of the Rhodes Memorial Lectures at Oxford in the autumn of 1936,

> The assumption of uniformity has much to be said in its favour. If the distribution were not uniform, it would either increase with distance, or decrease. But we would not expect to find a distribution in which the density increases with distance, symmetrically in all directions. Such a condition would imply that we occupy a unique position in the universe, analogous, in a sense, to the ancient conception of a central earth. The hypothesis cannot be disproved but it is unwelcome and would be accepted only as a last resort in order to save the phenomena. Therefore, we disregard this possibility and consider the alternative, namely, a distribution which thins out with distance.
>
> . . .
>
> The apparent departures from uniformity in the World Picture are fully compensated by the minimum possible corrections for redshifts on any interpretation. No margin is left for a thinning out. The true distribution must either be uniform or increase outward, leaving the observer in a unique position. But the unwelcome supposition of a favoured location must be avoided

> at all costs. Therefore, we accept the uniform distribution, and assume that space is sensibly transparent.[17]

First and foremost, it is clear from these words of Hubble that there are alternative interpretations of redshifts. I have done a simple text analysis of Hubble's three Rhodes Memorial Lectures. Interestingly, Hubble used the words "interpret" and "interpretation" as many as forty three times.

At least Hubble would agree that observations need to be interpreted with a particular "World Picture." In fact, Hubble gave the punchline in his lectures: "The fundamental question is the interpretation of redshifts."[18]

It is rare these days to hear about the alternative interpretations of redshifts. Nevertheless, a physicist in 2018 compiled as many as fifty nine different mechanisms that have been proposed over the years to explain the redshift.[19] Most of the explanations don't make sense and have problems of their own.

Another important feature of the universe is the density of galaxies. Hubble went on to say in his Rhodes Memorial Lectures:

> The departures from uniformity are positive; the numbers of nebulae increase faster than the volume of space through which they are scattered. Thus the density of the nebular distribution increases outwards, symmetrically in all directions, leaving the observer in a unique position. Such a favoured position, of course, is intolerable; moreover, it represents a discrepancy with the theory, because the theory postulates homogeneity. Therefore, in order to restore homogeneity, and to escape the horror of a unique position, the departures from uniformity, which are introduced by the recession factors, must be compensated by the second term representing effects of spatial curvature.
>
> There seems to be no other escape . . .[20]

What can we gather from Hubble's 1936 lectures? It seems that he abhorred the possibility of Earth being special. A "unique position" for Earth in the universe is abominable even today. The cosmic Copernican principle must be religiously obeyed at all costs.

The observed "departures" from uniformity in galaxy distribution has an unwelcome implication. It means that the Earth might be at the center of the universe. He had to either take the "positive" observation of

17. Hubble, "Observational," chapter 3 sect. 2.
18. Hubble, "Observational," chapter 2 sect. 12.
19. Marmet, "Interpretation," abstract.
20. Hubble, "Observational," chapter 3 sect. 6.

"departures" as it is, or to find a way to "compensate" it. "There seems to be no other escape." So, Hubble introduced "the second term representing effects of spatial curvature."

This is a classic example of how a worldview can affect the decision of a scientist in a scientific work. Whenever observations suggest a "unique position" of Earth, cosmologists will try to find ways to "compensate" them.

But as discussed in the next chapter, we see the alignment of Earth's elliptical orbit with the cosmic equator of the so-called Cosmic Microwave Background (CMB) map. This impossible coincidence suggests that there is something special about Earth's position in the universe. This alignment would likely be a horror to Hubble if he were still alive. No wonder cosmologists call it the "axis of evil."

In any case, Hubble concluded his lectures by saying,

> Two pictures of the universe are sharply drawn. Observations, at the moment, seem to favour one picture, but they do not rule out the other. We seem to face, as once before in the days of Copernicus, a choice between a small, finite universe, and a universe indefinitely large plus a new principle of nature.[21]

Let's unpack his somewhat cryptic conclusion of his three lectures. Hubble basically gave us two possibilities: either the "small"[22] and "finite" universe, or the infinitely large universe. The choice is between the Copernican principle and a "new principle" of Nature. These two options are mutually exclusive worldviews.

Even if Earth is at the center of the universe, it means nothing. Though God created everything with a focus on Earth, nowhere in the Bible does it say that the Earth occupies a special place in the universe. What God is concerned with is not even the Earth itself, which He had cursed because of Adam's sin. He will destroy the Earth in the future. Whether you believe it or not, the Bible in its totality tells us that God's concern is us—fallen humans, male and female, created in His image.

So, I am rather surprised by the trepidation and unnecessary intolerance to interpretations incompatible with the cosmic Copernican principle. For sticking to this principle, Hubble told us that he had to introduce an additional parameter to "compensate" the observation data. The price he paid is the Occam razor. Such intolerance makes the Big Bang model less convincing.

21. Hubble, "Observational," summary.

22. According to the more accurate measurements, under the Big Bang paradigm, Hubble grossly underestimated the size and age of the universe.

Thus, we see that the Big Bang model requires redshifts to be interpreted as space itself is expanding. No one can live forever to observe directly the movement of a galaxy in the universe. We are not to confuse this movement with the apparent motion in the sky due to Earth's rotation. Our existence on this planet is not long enough to accumulate data over millions of years. We only have written history *far* less than ten thousand years.

5.2.3 Observations are Incompatible with Isotropy

A new study using observations by an X-ray telescope is challenging the assumption of uniform expansion. The researchers observed hundreds of galaxy clusters. They found that some galaxy clusters appear to be moving away from us faster in some directions than others.[23] This behavior is not allowed in the standard Big Bang model. It requires space to expand evenly in every direction.

The X-ray scientists also compared their findings with studies from other research groups that have found indications of a lack of isotropy using different techniques. They established agreement on the direction of the lowest expansion rate. Once again, we see that a fundamental pillar of the Big Bang cosmology—isotropy—is cracking.

To rescue the Big Bang model, the X-ray cosmologists provided two explanations for the observed asymmetry. One of the explanations is the possibility that dark energy is not uniform, stronger in some parts of the universe than others. As a result, we have different expansion rates. This explanation is a speculation. Nobody knows what dark energy is. Using the unknown to explain the puzzle is acceptable in cosmology.

What is the takeaway? It seems that the universe is much more complex than what the Big Bang theorists have anticipated.

The redshift is an irrefutable reality established by observations at present time. The microwave radiation in the background too is measured with higher and higher precision today. These are prime examples of observational science.

But as Hubble had said, "The fundamental question is the interpretation." For the last seventy years, Big Bang cosmologists dictate their standard interpretation as orthodoxy. It is rather similar to the period of history when only the religious leaders of the Roman Catholic Church had the prerogative to interpret the Bible.

23. Migkas et al., "Probing," abstract.

Given the scientific cracks that some astronomers have brought to our attention, it isn't improbable that in the future, an overhaul of the standard Big Bang cosmology will happen.

5.3 By Faith, Extrapolate

Ultimately, what is the problem with Big Bang? The short answer is that scientists working in these fields need to extrapolate back to dark deep time, which is unknown. Let me ask a question. "What was going on seven thousand years ago?" Do you know? Most likely, nobody knows. It is the same with dark deep time in the past.

5.3.1 Thought Experiment on a Candle

To illustrate the crux of the problem, suppose you enter a room, and you see a lit candle. Out of curiosity, you ask a simple question: "For how long has the candle been burning?" You are trying to make some guesses about what have happened to the candle before you enter the room.

As a scientist, you immediately measure the length of the candle, say ten inches. An hour later, you measure the length again. You find that the candle's length is shortened by an amount. Let say it's one inch. You deduce that the rate of burning is an inch per hour. At this rate, and given the remaining length of the candle, which is nine inches now, you predict that the candle will burn out completely in nine hours. So far so good. Everything is scientific and rational.

But these measurements don't answer the original question you have posed. What have happened before you enter the room? So, you reason that since the candle is getting shorter and shorter, it must have been longer in the past. So, you make an assumption. You assume that the original length of the candle before it was lit must not be taller than the ceiling.

The candle is placed on a desk. As you have no ladder to reach the ceiling, you then perform a modeling of the ceiling height based on the height of the desk instead. From your modeling and measurement of the desk height, you *infer* that the length of the original candle should be, say, 148 inches. From this estimate and given that the candle was ten-inch long when your entered the room, you say that the candle has been burning for 138 hours. This part is an extrapolation back in time. It is historical science.

The analysis of candle burning is exactly what the Big Bang cosmologists are doing. The height of the ceiling puts a limit on how long the candle can be. This limited height corresponds to Friedmann's scale factor, which is limited to be greater than zero. The observable desk corresponds to the observable redshift. And your model corresponds to the solution of the Friedmann equations. Rewinding the clock, you infer that the candle has been burning for 138 hours, and the universe has been around for 13.8 billion years. This is the method of historical science.

But you didn't actually observe the original candle length. Did you? You came into the room when the candle is already burning. Likewise, the universe is already here before cosmologists do the study. Scientists don't and can't measure 13.8 billion years directly. This age is inferred from the model.

Some cosmologists are so bought into the Big Bang model that they stop thinking about running the video backward anymore. They simply fast-*backward* to the "beginning" and start the Big Bang as if it had already happened in history. Big Bang must be taken for granted. How else can any academic cosmologist publish papers in top journals?

But did you detect a fundamental problem in this practice of extrapolation into dark deep time? Let us return to the candle thought experiment again. The crucial bit is that the history of the candle is a result of the model with the assumption that it can't exceed the ceiling. There is no direct observation of the past candle when it is thought to be longer than what it is now. Since it is about the past, we can't travel back in time to observe the candle. Since it is the only candle in the room, you can't perform repeatable experiments either.

Without repeatability, such modeling of the past is not really science. It appears to be science because of your mathematical modeling and observations of the height of the desk and so on. But it can't be science because nobody can verify or falsify the past candle produced by the model. Similarly, nobody can verify or falsify big evolution produced by the Big Bang model.

Many unknown unknowns are possible in the past. For example, it could well be that an hour before you enter the room, someone lit the candle. In that case, the candle was burning for just an hour before your arrival. The age of the burning candle is just an hour, not the inferred 138 hours.

5.3.2 Friedmann's Simplest Cosmology

I shall now describe the cosmology depicted in Friedmann's 1922 paper. You can see the parallel with the candle illustration.

As a recap, the main assumptions required by Friedmann to solve Einstein's field equations are as follows:

A. Cosmic Copernican principle: isotropic, i.e., no particular preferred direction and location

B. Homogeneity: uniform; no big void and big lump in the universe at the large scale

C. Perfect fluid: homogeneous pressure

The cosmic Copernican principle simplifies Einstein's field equations dramatically. The three-dimensional space is described by just a scale factor, which is a function of time. By convention, we can set the time now as 0 and the scale factor as one.

The motivation for the other two assumptions is to grossly simplify the right-hand side of Einstein's field equations. These two assumptions make the distribution of matter and energy ideal for modeling. Without these three assumptions, Einstein's field equations can't be easily solved.

In fact, all the three assumptions fail to hold. Slowly but surely, peers-reviewed papers have accumulated a wealth of evidence to suggest that these three assumptions are questionable. As discussed earlier, they include the observations of anisotropy from X-ray astronomy, the cosmic axis of evil in CMB data, and the presence of cold spots in the supposed uniform distribution.

Nevertheless, for the time being, let us pretend that these three assumptions are okay. Friedmann derived two differential equations for the scale factor to follow religiously.

The simplest solution of Friedmann's equations is one for which the perfect fluid has only the mass-energy density and the pressure is zero. In this so-called matter-dominated case, Friedmann's two equations reduce to just one. This one equation can be easily solved by high-school calculus.

The solution is plotted in Figure 5.2. It's a simple function of time t raised to the power of 2/3, i.e., $t^{2/3}$. Being a concave function, the rate at which Friedmann's universe expands is forecast to be decelerating.

How much will the universe expand one year from now? This simple function allows us to address this question. We do the calculation. The answer is seven billionth of a percent. It is a negligible tiny amount. We can't measure this tiny increment with the best telescopes today.

Next, for the universe to expand by one percent, this simple Friedmann solution predicts that it will take more than 138 million years. But again, since no one can live to millions of years old, no one can scientifically test this prediction about the future of the universe.

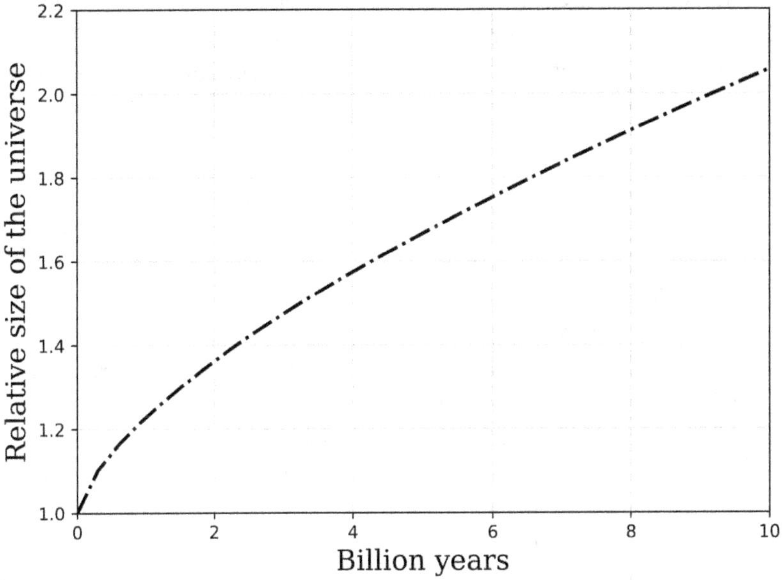

Figure 5.2. Simplest solution of Friedmann's differential equations of the scale factor, which represents the size of the universe relative to today's size.

Nothing is practically testable.

You can see that indeed there is a parallel with the candle thought experiment. Having determined the rate of burning, you then use a model to predict the length of the candle one hour from now, deterministically. Your model corresponds to this simple Friedmann solution (Figure 5.2), and the rate of burning represents the Hubble constant.

In a nutshell, cosmologists extrapolate into the future, and by faith, cosmologists believe that the universe will behave exactly in the way that the model dictates.

This forward prediction is not really a problem for your candle experiment, because you live *outside* the candle, and you live long enough to verify whether your prediction of nine hours for the candle to burn out is correct or not.

But the same can't be said for the universe. We live *inside* it and the universe is so vast that a tiny change over a year predicted by the simplest Friedmann solution is no different from noise in the observation. It is buried in the noise and thus unobservable. To obtain a detectable expansion in the future from now requires millions of years.

Imagine you are a sentient molecule in the candle. How can you tell that the candle is burning or not? In the same way, how do you really know whether the prediction of modern cosmology is true? This practical limitation of verifiability is always the destiny of cosmology.

5.3.3 Why Don't We See Galaxies Expanding?

To illustrate cosmologists' assertion that it is space that is expanding, we can perform a simple experiment as follows. Let's draw a few small circles on a half-blown balloon. We can observe that although the circles obviously can't move by themselves, every circle is moving further apart from each other when the balloon is being pumped with air.

The surface of the balloon represents space. When space expands, the wavelength of the light from the galaxy gets "stretched" and becomes longer as it travels through the expanding space to reach our telescopes. This is the mainstream interpretation of redshift.

What cosmologists don't tell you in this illustration is what happens to the circles. Each of them represents a galaxy. If the space expands uniformly, the small circles should become bigger as well. You can observe this expected outcome in your balloon experiment. You can also find some circles turn into ellipses.

As the universe expands, why don't galaxies get stretched out? According to an astronomer, the reason is as follows:

> There is enough matter in a galaxy, that the matter within the galaxy is not affected by the expansion of the universe. You can think of this as the gravity of the galaxy holding it together, but really it's more fundamental than that. The rate of the expansion of the universe depends on the amount of matter (and dark energy) in the universe. If you just consider a tiny fraction of the universe which just includes a galaxy and total the matter in that region, it's more than enough to have already stopped the expansion in that region.[24]

So the space within a galaxy is not expanding because it has enough matter to stop the expansion. We shall apply this dictum in the following subsection.

Indeed, at the local scale, the space is practically not expanding. Otherwise, we should see the Sun getting further away when the vast space between the Sun and the Earth expands. Our planet will get colder and colder, and the time it takes to orbit the Sun a full cycle should become

24. Masters, "Universe," para 1.

longer than one year. But the observed regularity of our solar system doesn't support the idea that it is expanding. In fact, we don't see any galaxy itself expanding either.

5.3.4 Problems of Backward Extrapolation

Now, what Big Bang theorists seldom tell us is that they are always making an extrapolation assumption by winding back the clock. Whenever they talk about the Big Bang from a "cosmic egg" or "nothing," they need to reverse the arrow of time. But nobody can observe extrapolation.

The Friedmann differential equations are deterministic. In principle, there is no physical constraint for reversing the time as thermodynamics is not integrated into Einstein's theory of general relativity. But if it was integrated, then the second law of thermodynamics would forbid time to reverse.[25] In less technical terms, it is something you already know. Nobody can return to yesterday.

The backward "forecast" with negative time, backcast in short, is essentially no different from the forward forecast discussed earlier. All the concerns of being unobservable for forecast apply to backcast too.

If we wind the clock back to a billion years from now, according to the simplest Friedmann solution, the scale factor would be 0.926. In other words, the universe a billion years ago would have been smaller than the present size by about 7.4 percent. On average, the galaxies would have been closer to each other by this percentage. But this percentage for the past is an unobservable speculation. Obviously, no scientist can test this backcast value.

At negative 4.5 billion years, when the solar system was said to be forming, the scale factor would have been 0.639. The universe then would be much more congested. By Newton's inverse square law of gravity, the gravitational pull of every galaxy on each other would be 2.4 times stronger than it is today.

As we continue to move backward, there would come a time when all the galaxies were very close to each other. All the galaxies were so close that the gravitational interaction becomes a lot stronger.

Now, let's go "back to the future" from this past universe of almost overlapping galaxies. But then we have a problem. According to the dictum

25. The second law of thermodynamics is valid for closed systems only. A closed system has no interaction with the outside world. Although we don't and we can't know what's "outside the universe," it's appropriate to assume that the universe is a closed system.

that the galaxy is not stretched by universal expansion, the overlapping galaxies can't expand because we now have "more than enough" matter to stop the expansion. It means that the universe couldn't expand from that time onward anymore. The universe would remain static. An "accident" has happened.

It seems that we have a fundamental contradiction when we extrapolate time back to the past. This contradiction is the result of Strogatz's "accidents" when we use the "fundamentally naive" calculus "to reconstruct the past."

Another possibility is that the massive black hole in each galaxy would merge together when the clock is reversed, ultimately forming the Great Wall of Big Black Hole. When all matter of the universe is inside the Big Black Hole, Einstein's field equations are no longer applicable. By extension, the Friedmann model stops working, and it is impossible to wind back time all the way to the supposed moment of Big Bang.

Obviously, I can't prove that the Great Wall of Big Black Hole did happen in the past. But neither can this scenario be ruled out as theoretically impossible.

As a summary, the Friedmann equations provide a *theoretical* possibility to wind back time, moment by moment. The argument is that if the universe is expanding now as predicted by Friedmann's cosmology, it must have been smaller in the past. So let's follow the Friedmann solution carefully, frame by frame, like a video playing backward.

Big Bang proponents usually disregard this frame-by-frame analysis. What would happen if the galaxies were almost on top of each other? Cosmologists need to take a quantum leap of faith over 13.8 billion years to the purported beginning of Big Bang. As we wind back the time with Friedmann's solution, we might get the Big Black Hole formed by the mergers of galaxies. Then the Friedmann trajectory fails and we can't wind back anymore. Time ceases to exist in a black hole. Accidents can happen.

Extrapolation to either the future or the past with a model needs faith. This is because scientists have no way to verify whether the universe would really behave in the way prescribed by the solution of the Friedmann equations. Cosmologists have to believe, by faith, that over billions of years, not a single surprise springs up along the way. All the physics known today is enough to dismiss the unknown unknowns. This assumption is quite preposterous as it implies that all the most fundamental laws of physics have already been found. The end.

But Nature always surprises the discoverers. A discovery is not obvious before it's discovered.

For the time being, let's assume there are no surprises in traveling back in time. In the simplest matter-dominated Friedmann solution, the time when the scale factor equals zero is 2/3 the reciprocal of the Hubble constant. A simple calculation gives us 9.2 billion years for the "age" of the universe. This calculated age is unacceptable because astronomers have observed galaxies much older than it.

Now what? How to make the universe older?

As it turns out, fudge factors are added to make the universe older. Cosmologists introduced dark matter and dark energy to the Big Bang model.[26] In cosmology, the word "dark" is a secret code for "unknown."

The standard model of Big Bang has six parameters. Using these parameters, scientists can also calculate the composition of the universe. Intriguingly, dark energy has the lion share of 68.3 percent, and dark matter accounts for 26.8 percent. The "normal" matter we see in the universe is just 4.9 percent.[27]

So, with more than ninety five percent of the current universe still unknown, can we rationally think that cosmologists have the final version of the cosmic history from Big Bang?

Is it possible that Big Bang is a wrong model?

26. Ade et al., "Planck 2015," 8.

27. Planck Collaboration, "Planck's," para. 1.

6. Is there a "New Genesis"?

> Everyone who hears these words of Mine and does not act on them, will be like a foolish man who built his house on the sand. The rain fell, and the floods came, and the winds blew and slammed against that house; and it fell—and great was its fall.[1]

6.1 Theoretical Physics and Surprises

THEORETICAL PHYSICS OF THE kind carried out by Schrödinger and Dirac is much harder to explain in simple terms. With neither pictures, data, nor concrete images to show, I can only speak in general terms. So I shall begin by giving a narrative of my journey in theoretical physics.

Anyone who had done some theoretical physics in mid 1980s and 1990s would know a theoretical physicist by the name of Edward Witten. He is so smart that I heard from someone that Witten has two brains, one of which is under his chin. I hope this is not derogatory, as I have a lot of respect for him.

Almost all of Witten's papers written during this period became a subfield of string theory. Many theorists and even some mathematicians, especially those in Witten's inner circle, followed his works closely. They mopped up certain details or further extended Witten's new papers.

Even this kind of sub-contractor works is admirable, given the amount of math one must master and apply with insights. Witten's contemporaries would probably agree that if Witten were to claim to be number two, no one would be foolish enough to claim to be number three, let alone number one.

His math is so difficult yet insightful. In 1990, Witten won the Fields Medal, the most coveted honor of any mathematician. Every four years, the Fields medals are awarded up to four mathematicians who have made outstanding contributions. They must be under forty years of age. At an average

1. Matthew 7:26–27.

of (less than) one math expert per year, the Fields medal is arguably more difficult to win compared to the Nobel prize for physics.

What is amazing about Witten is that his bachelor's degree is in history with a minor in linguistics. Although his father was also a theoretical physicist, Witten probably learned physics and math on his own. Witten's math is so multi-faceted and significant that the 1966 Fields medalist Michael Atiyah had this to say about him:

> In all this large and exciting field, which involves many of the leading physicists and mathematicians in the world, Edward Witten stands out clearly as the most influential and dominating figure. Although he is definitely a physicist (as his list of publications clearly shows) his command of mathematics is rivalled by few mathematicians, and his ability to interpret physical ideas in mathematical form is quite unique. Time and again he has surprised the mathematical community by a brilliant application of physical insight leading to new and deep mathematical theorems.[2]

To do theoretical physics, you must learn a lot of math and also how to apply it. Theoretical physics is applied math; we all know that. But for the first time in history, Witten turned the table around. He applied theoretical physics to solve some deep problems in pure math. For once in a few thousand years perhaps, math is applied theoretical physics.

Witten may not be as famous compared to Hawking. But Witten's prolific discoveries in the mathematical structure of quantum field theory "have opened new paths in all areas of quantum physics. Witten is the originator of M-theory, which united five competing string theory models. This result kicked off intense interest and rapid developments in the field, leading to a boom known as the second superstring revolution."[3]

Now, back in early 1990s, I used to hang out with two PhD students (Phil and Singh) and discussed physics over lunch. I had a working space at the National University of Singapore (NUS). The science library was nearby. I often went there to look out for the latest papers published in journals such as *Physics Letters B* and *Nuclear Physics B*. I would make a mental note of the new papers and the latest developments in string theory.

However, by the time I came to learn about Witten's key idea and the math needed to follow one of the trails he had blazed, many follow-up papers had already appeared on an online pre-print server called arxiv.org

2. Atiyah, "Work," 514.

3. The 2016 inaugural American Physical Society (APS) Medal was awarded to Witten. This quote is APS's citation of his achievements. See Conover, "Edward," para. 5.

today. The learning curve was very steep. But somehow I wasn't deterred by the mathematical difficulties.

Rather than getting bogged down by mathematical rigor, I learned the essentials of math quite extensively. The areas of math I picked up includes Lie group, Kac-Moody algebra, braid group, representation theory, differential geometry, Riemann surfaces, and so on. These different areas of math are applied in theoretical physics. We need them to describe the geometry and the symmetries of the theory. Symmetries are very important in physics and they are related to transformations. For example, if we transform a coordinate system into another, the physics should remain the same. This is described as translational symmetry.

I was particularly interested in topological invariance. Topology is the study of an object's proprieties that are unaffected or invariant when it is stretched or squeezed.

Imagine for example the concept of winding number. With reference to a particular point on a two-dimensional plane, you draw a circle around it. The winding number is said to be one round, or simply, one. The circle need not be perfectly drawn. It can be of any shape so long as the starting point and the ending point match exactly to complete one round. In other words, winding number is a topological quantity. No matter how badly drawn the circles are, when the starting and ending points coincide, you can go around a point on a flat plane many times to obtain a larger winding number.

While studying Dirac's function, I came to realize that I could use one of the "representations" of Dirac's function to characterize the topological properties of quantum Hall effect. The Dirac function is quite a strange construct. It is zero everywhere except at a point, where the function blows up to infinity. So long as this defining property is satisfied, the Dirac function can be expressed or "represented" in various forms.

Each form is a representation of Dirac's function. I found a certain representation useful for the theory of quantum Hall effects. It can incorporate the winding number. This idea was the starting point that blossomed into four published papers for my PhD.

What then is a quantum Hall effect? To explain it, we have to start with the "classical" Hall effect. In 1879, Edwin Hall performed experiments to study the electric current flowing in a sheet of metal when a strong magnet was put near it. He discovered that the electric current was affected by the magnet, giving rise to a voltage difference perpendicular to the direction of the electric current. This voltage is dubbed the Hall voltage.

Intuitively, the Hall effect can be understood as a side-way movement of the electrons under the influence of a strong magnet. The movement of electrons is electric current. When the Hall voltage is divided by this side-way current, we obtain the Hall resistance, which is measured in the unit of ohm.

Real science is indifferent to history. Whether the experiments are performed in the nineteenth century or the twenty first century, Nature will provide the same answer. When we carry out the experiments which Hall did, we will obtain the same results.

The operational science of the Hall effect is applied as sensors in a wide range of applications. For example, in the automobile industry, Hall effect sensors are applied to sense the positions of seat and safety belt for airbag control. Do you know that Hall effect sensors are in your washing machine? They help it maintain balance during washing.

The Hall effect is caused by the presence of a very strong magnet. What will happen to the Hall effect when an ultra-thin sheet of conducting material is cooled to a very low temperature? Strange things unseen before occur. Empirically, in 1980, Klaus von Klitzing discovered that at low temperature, electrical resistance of the Hall effect varies not by a smooth scale but in a series of precise jumps.

Amazingly, the jump amplitude depends only on two fundamental constants of Nature. We multiply the charge of an electron with itself, and then use the resulting square to divide the Planck constant. We obtain a value of about twenty six thousand ohm. This is known as the von Klitzing constant. The integer quantum Hall effect is the manifestation of the discrete multiples of this constant.

This is the first surprise. No theory hitherto predicted that a system of interacting electrons in a magnetic field would produce integer quantum Hall states. But not long after the discovery, we can explain the integer quantum Hall effect with quantum physics.

The second surprise came when the quantum states were found to be a fraction instead, such as 1/3, which was once thought to be impossible. The new physics of this observed phenomenon is called the fractional quantum Hall effect. A theorist by the name of Robert Laughlin provided a quantum theory to explain the new experimental results. Laughlin's theory is highly original. However, it works only for fractions of odd-number denominators.

It turns out that Nature vouchsafed more pleasant surprises. It also allows fractional quantum Hall effect with even-number denominator. What I did was to develop a topological quantum field theory to classify all the fractional quantum Hall effects with Laughlin's model as a special case.

Furthermore, my extended framework allowed me to predict a certain fractional possibility that had not yet been discovered.[4] I managed to publish four papers in three years.

My main contribution is to account for fractional quantum Hall effects that can't be explained by Laughlin's theory. Additional ideas are needed to accommodate other fractions. Even so, in one of my published papers, I acknowledged that "it is not unfair to say that they all build upon the conceptual foundation of Laughlin's theory."[5]

So it came as no surprise to me that in 1998—four years after I was awarded the PhD—Laughlin won the Nobel prize. It was a joint award with Störmer and Tsui, who discovered the fractional quantum Hall effect in their lab.

The quantum Hall effects are real science. They have experimental verifiability, and their theoretical explanations are rock solid. They provide another confirmation that quantum mechanics is the correct fundamental framework to *describe* how Nature *operates* at the sub-atomic scale.

Conventional quantum computers are sensitive to thermal fluctuations. They need certain error correction algorithms to remove the unavoidable mistakes. On the other hand, the topological nature of the fractional quantum Hall effect means that it is robust against the inescapable imperfections in the materials and thermal noise even at low temperature. It remains to be seen whether topological quantum computing will become ubiquitous.

In research, sometimes you aim for something but eventually you get something else. I wanted to do string theory but eventually I was doing theoretical physics associated to repeatable experiments. I wanted to do particle physics but ended up with single-headed heavy meromyosin and fractional quantum Hall effect.

When I was pursuing a PhD in physics, I didn't know—never did it cross my mind—that Kepler and Galileo were grateful to God for allowing them to discover the afterglow in Nature that speaks of God's awesome glory and majesty. At the end of Acknowledgment in my PhD thesis, I wrote,

> As a praise of Him who made Nature so beautiful and full of surprises.[6]

4. Ting, "Topological," 103. Specifically, the fraction is 5/7. This fraction has been found after my PhD thesis.

5. Ting and Lai, "Spinning," 430.

6. Ting, "Topological," i.

6.2 Hubris

While studying and doing physics, I came to know of many geniuses, either through their works, or through sitting in a lecture room to learn from their lectures. Recalling a conversation with Professor Lai, my thesis supervisor, I heard him saying that a theoretical physicist by the name of Goldstone was so smart that he had done almost all the theoretical physics. He didn't bother to publish but people recognized his brilliance. If you would ask Goldstone a current conundrum of physics, he would pull out one of his many folders in one of his office cabinets and show you his analysis and proofs.

Such is the brilliance of some physicists, among whom was Stephen Hawking, an atheist. Surely nobody will dispute that Hawking was a genius. Though I have never met him, through his papers, I realize that Witten is the genius of geniuses.

Having recently spent some time reading the translated versions of their scientific works, I came to realize that Kepler and Galileo were geniuses too. In spite of their intellectual prowess to probe Nature, they genuinely believed in God of the Bible. They held the view that the Scripture is the inspired Word of God. They honored God through their scientific inquiries.

Add to them their successor Newton, who certainly had done a lot of Bible study. It was said that in terms of the number of words written, Newton wrote more on Christianity than on physics. He believed in God of the Bible. Nevertheless, it is an irony that despite his career affiliation to Cambridge University's Trinity College, Newton's theology was strictly not trinitarian.

Scientists these days are far removed from the milieu of Kepler, Galileo, and Newton. Since the later part of eighteenth century, the climate has transformed completely from conducting scientific research to glorify God of the Bible. Not only is God no longer revered anymore, God is dead. The reverent fear of God seems to be buried in the cacophony of the "free spirits" in the air (*Zeitgeist*). In "We fearless ones" from *Gay Science*, Friedrich Nietzsche in 1882 wrote about "What our Cheerfulness Signifies" as follows.

> The greatest recent event—that "God is dead," that the belief in the Christian god has become unbelievable—is already beginning to cast its first shadows over Europe. For the few at least, whose eyes—the suspicion in whose eyes is strong and subtle enough for this spectacle. Some sun seems to have set and some ancient and profound trust has been turned into doubt; to them our old world must appear daily more like evening, more mistrustful, stranger, "older."

. . .

> In fact, we philosophers and "free spirits" feel, when we hear the news that "old god is dead," as if a new dawn shone on us; our hearts overflow with gratitude, amazement, premonitions, expectation . . . the sea, *our* sea, lies open again; perhaps there has never yet been such an "open sea."[7]

Before Nietzsche finally publicized the obituary of Jesus, there was already an unspoken but implicitly corroborated philosophical movement leading to it. Most scientists since then wanted to emancipate themselves from Christianity by categorically denying God as the Giver of natural laws.

It's not quite clear what "our sea" in Nietzsche's prose quoted above may be interpreted. But if it means the world of knowledge, i.e., science, we recall Newton once famously said about "the great ocean of truth":

> I do not know what I may appear to the world, but to myself I seem to have been only like a boy playing on the seashore, and diverting myself in now and then finding a smoother pebble or a prettier shell than ordinary, whilst the great ocean of truth lay all undiscovered before me.[8]

Nietzsche's philosophy is a big contrast to Newton's, as far as the east is from the west. Intriguingly, Newton wrote a "short Schem" of the true religion as follows.

> Religion is partly fundamental & immutable partly circumstantial & mutable. The first was the Religion of Adam, Enoch, Noah, Abraham Moses Christ & all the saints & consists of two parts our duty towards God & our duty towards man or piety & righteousness, piety which I will here call Godliness & Humanity.
>
> Of Godliness.
>
> Godliness consists in the knowledge love & worship of God, Humanity in love, righteousness & good offices towards man. Thou shall love the Lord thy God with all thy heart & with all thy Soul & with all thy mind: this is the first & great commandment & the second is like unto it, Thou shalt love thy neighbour as thy self. On these two commandments hang all the law & the Prophets. Mat. 22. The first is enjoyned in the four first commandments of the Decalogue & the second in the six last.[9]

7. Nietzsche, *Gay*, 279–280.
8. Spence, *Observations*, 158–159.
9. Newton, "Short," para. 1–2.

It is intriguing that a giant of science interpreted Genesis literally. He spoke about Adam, Enoch, and Noah in the same breath as he spoke about Abraham, Moses, and Christ. Newton didn't seem to consider that they were not individuals. Newton didn't look at them as myths or allegories, i.e., not real history.

Doubting the historicity of Christ, for better or for worst, is a form of apostasy and atheism. Newton's understanding of God is the One who gave the moral laws. Quoting what Jesus said in Matthew 22:37–40, Newton noted that Jesus had connected the two great commandments with the ten commandments. It is accurate to suggest that Newton had a great insight on the moral laws from God.

Robert Iliffe is a historian of science and director of the Newton Project. According to him, Newton's Bible study led him to the conclusion that "the orthodox notion of the Trinity was a fiction that was invented in the early fourth century and subsequently promoted by servants of the devil."[10] However, in the later part of Newton's life, "there was little evidence of the passion, or indeed, of the specific denunciation of orthodox doctrine that permeated his early works."[11]

Newton went on in his "Schem" to express his view about atheism and some of his reasons for his strong objection:

> Of Atheism
>
> Opposite to the first is Atheism in profession & Idolatry in practise. Atheism is so senseless & odious to mankind that it never had many professors. Can it be by accident that all birds beasts & men have their right side & left side alike shaped (except in their bowells) & just two eyes & no more on either side the face & just two ears on either side the head & a nose with two holes & no more between the eyes & one mouth under the nose & either two fore leggs or two wings or two arms on the sholders & two leggs on the hipps one on either side & no more? . . . Did blind chance know that there was light & what was its refraction & fit the eys of all creatures after the most curious manner to make use of it?[12]

It is also intriguing that Newton's objection to and his description of atheism as "so senseless and odious to mankind" is followed by asking many whys of biology. Newton probably would be against the notion of evolution by "blind chance."

10. Iliffe, "Newton's," para. 2.
11. Iliffe, "Newton's," para. 4.
12. Newton, "Short," para. 3.

Though separated by centuries, it would be enormously interesting if a hypothetical debate between Sir Isaac Newton and Mr. Charles Darwin were to take place. Incidentally, both were buried near each other in the north aisle of the nave of Westminster Abbey. This is a special place reserved for the burials of extraordinary British subjects. One died as a believer of God of the Bible, the other as an apostate.

Many atheists claim that atheism is a lack of belief in gods. Nevertheless, logically, either there are gods or no god. A *lack* of belief in something implicitly implies that that something exists. For example, we normally say, "Zeus doesn't exist." We don't say we lack a belief in Zeus, do we? In no uncertain terms, Newton stated that atheism is idolatry in practice.

In the context of Western civilization, which is the dominant one at present, atheism is a rejection of God of the Bible. It is still puzzling why Newton said that practicing atheists are idolaters.

Most likely, the idol here isn't in the traditional form of crafted wood or stone. Rather, it could be mankind's intellect and abilities, much like those who were building the tower of Babel to reach unto heaven. In short, consciously or subconsciously, we all want to beat God. Each of us wants to do it "my way." This is the original sin—the origin of all sins.

6.3 "New Genesis" and Big Bang

Now, once the train bound for secularism has departed the depot of theism, it gathers steam and speeds toward atheism. The restraint for pranks and outright mockeries has all but gone. For example, about the origin of the universe, some atheistic physicists have completely blotted out any last trace of the reverent fear of God. To many people, there is no God, and pious devotion to God is foolishness demonstrative.

In an "informal" autobiography, atheist George Gamow attributed his negative "attitude" toward the steady-state universe as a possible reason for not being invited to attend a prestigious conference. It was the 1958 Solvay Congress on Cosmology. In that era, steady-state universe was still the dominant theory.

Gamow was a creative theorist. He expressed his disappointment by replacing Genesis chapter 1 with his "New Genesis" in Exhibit 1.[13]

13. Gamow, *World*, 127.

New Genesis

In the beginning God created radiation and ylem. And ylem was without shape or number, and the nucleons were rushing madly over the face of the deep.

And God said: "Let there be mass two." And there was mass two. And God saw deuterium, and it was good.

And God said: "Let there be mass three." And there was mass three. And God saw tritium and tralphium, and they were good. And God continued to call number after number until He came to transuranium elements. But when He looked back on his work He found that it was not good. In the excitement of counting, He missed calling for mass five and so, naturally, no heavier elements could have been formed.

God was very much disappointed, and wanted first to contract the Universe again, and to start all over from the beginning. But it would be much too simple. Thus, being almighty, God decided to correct His mistake in a most impossible way.

And God said: "Let there be Hoyle." And there was Hoyle. And God looked at Hoyle . . . and told him to make heavy elements in any way he pleased.

And Hoyle decided to make heavy elements in stars, and to spread them around by supernovae explosions. But in doing so he had to obtain the same abundance curve which would have resulted from nucleosynthesis in ylem, if God would not have forgotten to call for mass five.

And so, with the help of God, Hoyle made heavy elements in this way, but it was so complicated that nowadays neither Hoyle, nor God, nor anybody else can figure out exactly how it was done.

Amen.

Exhibit 1. George Gamow's "New Genesis."

Gamow was a big evolutionist and a prolific writer. He authored many popular books on physics. After the Russian Revolution, Gamow underwent his doctoral program at the Leningrad University.[14] From 1923, he was under the tutelage of Alexander Friedmann. Regrettably, Friedmann died two years later.

14. The university was renamed to Saint Petersburg State University after the dissolution of the Soviet Union in 1991.

It was Friedmann who provided the earliest prime mover of what today is known as the Big Bang model. As an application of his equations, Friedmann wrote in his 1922 paper that the expanding universe could be ten billion years old.[15] He interpreted this age as "the time that has elapsed from the moment when the space was a point."[16]

Naturally, influenced by his professor, Gamow was an early advocate and developer of Lemaître's evolutionary model of cosmic origin. In the beginning the universe was a "cosmic egg." However, at that time, mainstream cosmologists were holding onto Aristotle's philosophy of an eternally existing and infinite universe. This worldview is consistent with the steady-state theory proposed by Fred Hoyle and his collaborators in 1948.

An article entitled "Sir Fred Hoyle" in the Britannica reads:

> Within the framework of Albert Einstein's theory of relativity, Hoyle formulated a mathematical basis for the steady-state theory, making the expansion of the universe and the creation of matter interdependent. Einstein assumed that the universe as a whole is static—i.e., its large-scale properties do not vary with time. This assumption, made before American astronomer Edwin Hubble's observational discovery of the expansion of the universe in 1929, was also natural; it was the simplest approach, as Aristotle had discovered, if one wished to avoid a discussion of a creation event. The notion that the universe on average is not only homogeneous and isotropic in space but also constant in time was philosophically attractive. Hoyle, Bondi, and Gold called it the perfect cosmological principle.[17]

Logically, an expanding universe can't be static. The two *philosophical* worldviews of the universe are mutually exclusive. That is, Gamow's support of the Big Bang model was incompatible with Hoyle's model. Both can't be correct at the same time. Only one of them can be right. Nevertheless, it is logically possible that both models are wrong.

It should be obvious that in naturalistic cosmology, philosophical assumptions are needed. Einstein wanted to make his cosmological theory consistent with a static universe with no beginning and no end. This is his philosophical presupposition. So he added a constant term to his general theory of relativity. Now you see, theoretical cosmology is dependent on worldviews.

15. Friedmann, "Curvature," last para.
16. Friedmann, "Curvature," footnote 11.
17. Britannica, "Fred," para. 2.

Intuitively, Einstein's constant term counteracts the attractive force of gravity. In a way, it serves as an anti-gravity device on spacetime.

But this addition of a constant term turns out to be a blunder after Edwin Hubble's observations suggest that the galaxies are moving away from us in all directions. Simply put, our universe is not static because space itself is expanding. Since the universe is expanding, any worry that the universe may collapse on itself becomes unwarranted.

In the "informal" autobiography, Gamow wrote about his discussion with Einstein as follows.

> Thus, Einstein's original gravity equation was correct, and changing it was a mistake. Much later, when I was discussing cosmological problems with Einstein, he remarked that the introduction of the cosmological term was the biggest blunder he ever made in his life.[18]

Now, what is the philosophical assumption of the Big Bang model? Well, I have mentioned it before—the cosmic Copernican principle. That is, the universe should always look the same from any location within it. Suppose you are able to travel to a planet in another galaxy. The principle says that you should also find that all galaxies are moving away from your new location. Always.

The cosmic Copernican principle underlying the Big Bang model is essentially a weaker version of the assumptions in Einstein's cosmology. For the Big Bang model, the universe on average is still assumed to be homogeneous and isotropic in space; only the assumption of being static over time is discarded. As a result, the universe is allowed to look different in the past from the one we now observe. Big Bang evolutionists can now tell us a *history* of how the universe came to be. A portion of this big evolution is Gamow's "New Genesis."

Today, the Big Bang model is widely accepted since the discovery of the so-called cosmic microwave background (CMB) in 1965. The CMB is dubbed the leftover afterglow of the Big Bang from an unimaginably hot and dense cosmic egg. It is *interpreted* as the leftover radiation from the early universe in the Big Bang model.

With two dark ingredients added to the original model of Friedmann's 1922 theory, the standard model of cosmology of today is called the ΛCDM model. The acronym ΛCDM stands for Lambda Cold Dark Matter. The first ingredient is cold dark matter. The second is dark energy expressed as Λ, which is the Greek capital letter "Lambda." Cosmologists

18. Gamow, *World*, 44.

use Λ to denote the cosmological constant term, which was ditched by Einstein after he realized his "biggest blunder" of his life.

In the standard model of cosmology, Einstein's "biggest blunder" was resurrected to fix the Big Bang model. Its new name is dark energy. Along with the cold dark matter, it was introduced to rescue the old Big Bang model.

But as we shall see later, the cosmological constant rears its ugly head again. This time round, its introduction leads to a discrepancy between the theory and the observation by a "catastrophic" amount unheard of in the brief history of time.

Now, the assumption of an isotropic and homogeneous universe with no special location and direction demands a perfectly isotropic and homogeneous distribution of CMB over the entire universe. To test this assumption, precision in the measurements of CMB can't be overemphasized.

A satellite named the Cosmic Background Explorer (COBE) was launched in 1989. Its mission was to measure the CMB with high precision. For reporting the evidence that supports the Big Bang model, two principal investigators of the COBE project received the Nobel Prize in physics in 2006. The COBE team was praised for being the game changer by turning cosmology into a "precision science."

Operating for nearly a decade since its launch in 2001, the Wilkinson Microwave Anisotropy Probe (WMAP) was the spacecraft succeeding COBE to measure the temperature of space. The European Planck spacecraft took over the job from 2009 through 2013. Observation of astronomy has entered the space age.

The microwave background is a science of repeatable observation. Like the redshift, its *interpretation* as the cosmic fossil of Big Bang is another matter altogether.

6.4 "New Genesis" and Nucleosynthesis

Against the backdrop of rapid development in modern cosmology, let us unpack Gamow's "New Genesis." First, we note that Gamow revived the word "ylem" used by the Greek philosopher Aristotle who put forth the concept of "fundamental matter." The word "ylem" in Gamow's "New Genesis" is a transliteration of the Greek word for matter. It was like the concept of a cosmic egg. Gamow and many physicists would say that from this egg emerged the protons and neutrons during the "first three minutes"[19] since the beginning of time as we know it.

19. Weinberg, *First*, chapter 5.

It would be nice if cosmologists have a video recording of Weinberg's "first three minutes." Then, Gamow's "New Genesis" could well be a valid interpretation of the first few sentences of Genesis chapter 1.

Some Christians are quick to concede. They rationalize that since the opening verses in Genesis seem to contain no moral teaching, it would be alright to accept cosmologists' *consensus* on how our universe evolved to what it is today by natural processes only. Well-meaning Christians embrace the Big Bang because this model has a beginning. On the surface, it seems consistent with Genesis 1:1.

But nothing could be further from the truth. In the subsequent sections, you will see new scientific observations cast doubt on the Big Bang model and its derivatives. These big evolution models are built on a shaky and patchy foundation on a dubious point.

The field of study that Gamow was alluding to in his "New Genesis" is called nucleosynthesis. It is the modern version of alchemy—a recipe for the transmutation of one atomic element into another one. Wouldn't it be wonderful to turn cheap copper into expensive gold? Chemistry is the reconfiguration of electrons. On the other hand, alchemy is the recombination of protons and neutrons. Chemistry is science. But alchemy is not science.

Probably "New Genesis" might be a joke, as Gamow might want to use it as an oblique swipe at Hoyle's theory of the universe in the steady state. It is rumored that Hoyle coined the term "Big Bang" to hurl his contempt for the competing model inspired by Friedmann's cosmology. The "New Genesis" could be Gamow's push-back against Hoyle's insult. Nevertheless, I doubt any God-fearing mortal would make a parody out of a sacred text, especially when it is about God's claim that He created everything.

I don't expect Gamow would understand the deep meaning of Genesis 1. Probably he was proud of his science, and he felt it was alright to paraphrase a lesser worldview other than his scientism. Some orthodox rabbis would likely find Gamow's "New Genesis" repugnant to the tune of the abomination of desolation.

Let's not judge Gamow's hubris of making himself superior to Moses too harshly. Gamow was no hypocrite. He honestly mentioned at the end of his "New Genesis" that nucleosynthesis was so complicated that no cosmologist "can figure out exactly how it was done." Welcome to historical science!

To explain this confession of ignorance, let's take a look at the present status of a simpler version of nucleosynthesis. It is called the Big Bang nucleosynthesis (BBN). It is a field in astrophysics that attempts to probe the origin of the lightest atoms without their electrons. Big evolutionists want to know how they might have been made in the first couple of minutes since the Big Bang.

Based on the latest observations collected by the Planck spacecraft, when we interpret the data in the framework of the standard model of Big Bang, we can affirm again the "concordance" of the abundance of deuterium (mass two). But for lithium, the inconsistency with the model is also confirmed. This is the well-known "lithium problem" in BBN. It has plagued big evolutionists for decades, and it remains an open question today.

Let's now unpack Gamow's "New Genesis" in light of BBN and a simplified nuclear network displayed in Figure 6.1. It is drawn according to a paper by Brian Fields, a BBN expert. Each box in Figure 6.1 contains a symbol for the element, along with a superscript indicating the total number of protons and neutrons that make up the nucleus of an atom. In physics, this total number is called the baryon number.

① $^1n \to {}^1p + e + \overline{\nu}$ ② $^1n\ (^1p, \gamma)\ ^2d$

③ $^2d\ (^2d, ^1p)\ ^3t$ ④ $^2d\ (^1p, \gamma)\ ^3\mathrm{He}$

⑤ $^2d\ (^2d, ^1n)\ ^3\mathrm{He}$ ⑥ $^3\mathrm{He}\ (^1n, ^1p)\ ^3t$

⑦ $^3t\ (^2d, ^1n)\ ^4\mathrm{He}$ ⑧ $^2d\ (^2d, \gamma)\ ^4\mathrm{He}$

⑨ $^3\mathrm{He}\ (^2d, ^1p)\ ^4\mathrm{He}$ ⑩ $^3t\ (^4\alpha, \gamma)\ ^7\mathrm{Li}$

⑪ $^4\mathrm{He}\ (^4\alpha, ^1n)\ ^7\mathrm{Be}$ ⑫ $^7\mathrm{Be}\ (^1n, ^1p)\ ^7\mathrm{Li}$

⑬ $^7\mathrm{Be} + {}^2d \to {}^9\mathrm{B}^*$ ⑭ $^7\mathrm{Be} + {}^3t \to {}^{10}\mathrm{B}^*$

⑮ $^7\mathrm{Be} + {}^3\mathrm{He} \to {}^{10}\mathrm{C}^*$

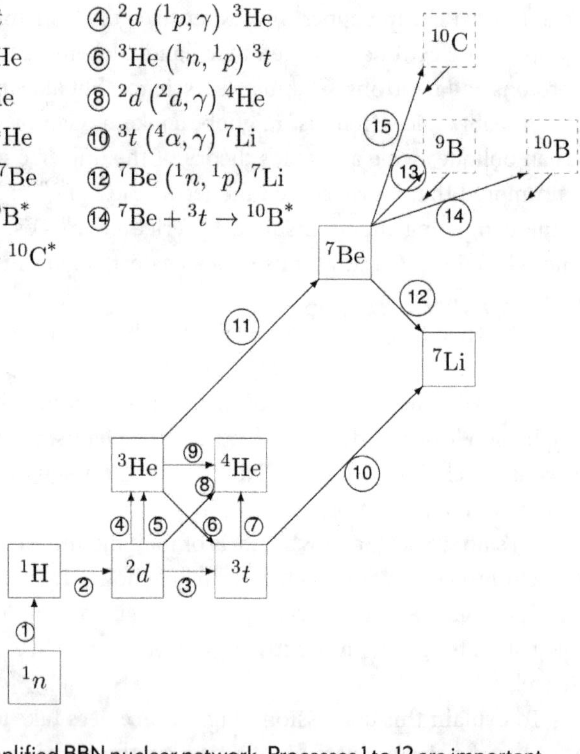

Figure 6.1. Simplified BBN nuclear network. Processes 1 to 12 are important nuclear reactions. The last three are proposed/tested new reactions.

Next, we need to refer to the periodic table of elements. Let's recall that every element is identified by the number of protons in the nucleus. This number is called the atomic number Z. For hydrogen (H), Z is 1; for helium (He), Z is 2; and for lithium (Li), Z is 3. The atomic numbers of beryllium (Be), boron (B), and carbon (C) are, respectively, 4, 5, and 6. Using the atomic number Z, you can figure out the number of neutrons. For example, lithium-7 with the symbol ^{7}Li in Figure 6.1 has four neutrons.

Every nuclear reaction in Figure 6.1 is generally represented as an incoming nucleus before the brackets and an outgoing nucleus after the brackets. Inside these two rounded brackets, there are also two symbols.

As anticipated, ^{1}p is the symbol for proton and ^{1}n for neutron. Moreover, γ is the usual symbol for photon. Its baryon number is zero, since it is a packet of electromagnetic energy. Photon is not matter.

The universal law of physics requires that all nuclear reactions don't change the baryon number, since nuclear reactions are merely the rearrangement of neutrons and protons. This is called the law of conservation of baryon number.

Take for example process 12 of Figure 6.1. Beryllium-7 reacts with a free neutron to produce lithium-7 and a free proton. When a proton or a neutron is outside of a nucleus, it is said to be free. Before the reaction, we have a total of four protons in beryllium-7, and four neutrons. Three of the neutrons are in beryllium-7. The remaining is the incoming free neutron. After the reaction, we also have four protons—three in lithium-7 and one proton becomes free. All the four neutrons are now in lithium-7, since the atomic number of lithium is 3. So we see that the baryon numbers before and after the nuclear transmutation exactly balance.

Now, despite the absence of direct evidence and scientific verifiability, we are told that during the first three minutes after the Big Bang, some neutrons ^{1}n underwent a nuclear reaction to become the hydrogen nucleus ^{1}H, which is simply a proton. This is process 1 in Figure 6.1, where an electron e and an anti-neutrino are also generated. Essentially, process 1 is the decay of free neutrons known as beta decay. Free neutron is unstable. On average, experimental measurements show that a free neutron lasts about fifteen minutes before beta decay occurs.

In nuclear physics, free neutron is puzzling. Different methods of measuring a free neutron's lifetime give varying estimates that are statistically different. Science is impossible if the property intrinsic to the object becomes dependent on the observation technique. No one has any clue why free neutron's lifetime is dependent on the measurement method. Since neutron's beta decay is one of the most important processes in the nuclear network, uncertainty in its rate of decay impacts the BBN analysis.

According to BBN, somehow a proton and a neutron meet each other. Through the nuclear reaction described as process 2 in Figure 6.1, they fuse into a nucleus of a proton and a neutron, which is deuterium. The mass of proton and that of neutron are approximately equal. So deuterium has two units of mass. In Gamow's "New Genesis," "Let there be mass two" refers to this nuclear fusion between a proton and a neutron.

Now, deuterium has only one proton. Thus, it is an isotope of hydrogen. An isotope of an element is its sibling. They have the same atomic number. Their difference is only the number of neutrons.

Mass three in Gamow's "New Genesis" corresponds to tritium. It is denoted by the symbol ^{3}t. Tritium is a rare and radioactive isotope of hydrogen. It is formed by process 3 in Figure 6.1, where two deuterium nuclei fuse together and release a free proton to form tritium.

Another possibility of mass 3 is tralphium, which is indicated by the symbol ^{3}He. It is a stable isotope of helium with two protons but only one neutron. As in Figure 6.1, tralphium is produced by processes 4 and 5. Process 4 suggests that deuterium somehow fuses with a free proton. Process 5 is the nuclear fusion of two deuterium nuclei. It produces ^{3}He and a free neutron.

It is important to note that, by process 6, tralphium can become the lighter tritium. This is the process where tralphium captures a free neutron and spits out a tritium along with a free proton. Nuclear reactions are not always making nuclei heavier. Like chemical reactions, reverse processes of producing lighter nuclei are plausible as well. If the reverse processes dominate, we will never obtain heavier elements through nucleosynthesis in the hot furnace of every star.

Helium, denoted by ^{4}He, is said to be produced in the BBN framework by three processes labeled 7, 8, and 9. It has mass 4—two protons and two neutrons.

Note that Figure 6.1 is a simplified version of the nuclear network, where the nuclear reactions that produce lithium of mass 6, i.e. ^{6}Li, are omitted. It can be produced by, for instance, the fusion of a deuterium ^{2}d and helium ^{4}He.

The more abundant lithium of mass 7 can be produced by process 10. In this nucleosynthesis, tritium fuses with an alpha particle, which is another way to refer to helium-4 nucleus, i.e., ^{4}He.

Intriguingly, there is no mass 5 in the nuclear network displayed in Figure 6.1. This is why in "New Genesis," Gamow revealed that God "missed calling for mass 5." So God had made a "mistake."

Many theologians would probably go ballistic at Gamow's theology: God could make mistakes and not knowing what was going on. According to BBN, the universe was in the beginning filled with hydrogen, deuterium, helium, and lithium. All other heavier elements are supposed to be produced by stars. When a star dies, its "star dust" of heavy elements is spread throughout the universe. Hoyle was the one who proposed this idea of star dust. That's why in Gamow's "New Genesis," we read that "Let there be Hoyle."

From this exposition of "New Genesis" in the context of BBN, we find that it isn't a rambling of Gamow. Rather, it's an accurate description of nucleosynthesis. Cosmologists believe that nucleosynthesis can explain the origin of atomic elements found on Earth. But this is a speculation. Nobody can historically trace the origin of nitrogen in our air to a particular dead star. By now, it's unobservable. Therefore, the claim of unknowable dead star as the origin of nitrogen is an untestable speculation.

In conclusion, cosmologists *assume* that every heavy atom in the universe was generated purely by natural processes of nuclear reactions that are going on in the stars. But can anyone say with one hundred percent certainty that uranium and all its isotopes found on Earth were the results of supernovae dark deep time ago? We see a puddle of water on the ground today. Does it always follow that it rained heavily many years ago?

6.5 Big Problems of Big Bang

If BBN could be directly verified to be true, then "New Genesis" would be more accurate than the original one in the Bible. We would need to make room for God to make mistakes and to be ignorant. This is a serious implication of accepting the Big Bang paradigm few Christians realize.

6.5.1 The Lithium Problem

But the story of "New Genesis" is blemished by the well-known lithium problem. Empirical observations of the so-called "primordial" lithium-7 in the stars are two to three times lesser than what the models of BBN suggest.

Theoretical physicists usually handwave away a factor of two in the initial formulation by intuition on the back of an envelope. Observational science, however, is a totally different ballgame altogether.

Compared to the data collected by COBE, the precision of WAMP data is much higher. Cyburt et al. reported in 2008 that the lithium problem worsens as data precision increases. The discrepancy between observations and the model becomes statistically more significant. Depending on the clustering of stars, the discrepancy is now 2.4 to 4.3 times. As a result of higher precision of data, the statistical significance becomes higher. It ranges from 4.2 sigma to 5.3 sigma.[20] Therefore, we can conclude that the observed amount of lithium-7 is well below the expected level of "concordance" prescribed by some BBN models.

It was indeed a bitter pill to swallow. Fields et al. concluded in their 2020 paper based on the most accurate Planck data as follows:

> As much as the concordance between BBN theory, light element abundance observations, and measurements of the CMB is impressive, the elephant in the room is the ^{7}Li abundance. There is a clear mismatch between the BBN predicted abundance and the abundance seen in metal-poor stars. We have largely ignored this discrepancy as it's not clear whether its resolutions lies in nuclear physics, beyond the Standard Model particle physics, the observational determination, or the stellar history of ^{7}Li. Obviously, a convincing resolution of this vexed problem would strengthen the power of BBN as the earliest probe of the Universe based on Standard Model physics.[21]

The cosmological lithium problem suggests that there is something "precisely" wrong about what happened during the "first three minutes." The elephant still doesn't want to leave the room. The most precise Planck data have confirmed the persistent discrepancy between BBN's predicted value and the current observations of lithium-7's spectral lines from the stars.

Hypothetically, what if the lithium problem was resolved? There is still a wide gap to claim that BBN actually happened dark deep time ago. A nuclear reaction that is experimentally observed today doesn't necessarily imply that it must have happened three minutes after the Big Bang. By construction, BBN was a historical event. If it really did happen, the historical BBN can't be observed today.

Take for example the decay of a free neutron. It is observed in the lab today. But to assert that the decay must have happened to jump start the nuclear network of BBN (Figure 6.1) is a speculation on history. We can't experiment with history.

Dark deep time is always involved when cosmologists extrapolate back in time from current observational science. Data and evidence gathered today don't speak for themselves. Models and worldviews are needed to interpret them.

20. Cyburt et al., "Update," 012.
21. Fields et al., "Big-Bang," last para.

6.5.2 Crisis in Cosmology

Lithium problem is not the only bitter pill. Another one is the value of the Hubble constant H_0.

It isn't an overstatement that H_0 is the most important measurement in cosmology. The Hubble constant H_0 today is the determinant of the age of the universe in the Big Bang paradigm. It also tells us about the pace at which the galaxies are receding from us.

By the cosmic Copernican principle, you are expected to see the same redshifts on another planet in another galaxy. It means that the Hubble constant is universal. The conundrum, however, is that the value of H_0 is dependent on the method employed to measure it.

I have earlier alluded to the puzzle surrounding free neutron's lifetime. Namely, a physical property intrinsic to the free neutron is dependent on the measuring method. It shouldn't happen in science. The measurements by different methods should give you the same value within the range of uncertainty.

But same thing happens to H_0. This time it is worst. We are now talking about the single most important and universal number in the grandest science of all—cosmology.

Specifically, a team led by the Nobel laureate Adam Riess used the Hubble telescope in space to measure the true brightness of rapidly pulsating stars known as cepheids. These stars serve as the candlesticks to measure distance far beyond our Milky Way.

Observation data from Riess's Supernovae H_0 for the Equation of State (SHOES) project led the team to narrow down the Hubble constant to 73.5, plus minus an uncertainty of 1.7. The SHOES method is a direct measurement of the current expansion rate at the present time.

But the CMB data collected by the Planck spacecraft gave a different result for H_0. When the data are interpreted in the standard ΛCDM Big Bang model, H_0 is 66.9, plus minus an uncertainty of 0.6.

It must be stressed that the Planck measurement is model dependent. The microwave background is interpreted as the left-over residue from about 300 to 400 thousand years after the Big Bang. The six parameters of the standard Big Bang model must be given the specific values.

Well, you may think that these two estimates—73.5 and 66.9—seem pretty close. It is a mere difference of 6.6, or about ten percent. The problem, however, is that the refined precision with which SHOES and Planck teams performed their measurements each gave a very narrow error bar. From the perspective of the SHOES estimate, Planck's estimation of H_0 is about 4 sigma smaller. Conversely, with respect to Planck's estimate, SHOES's H_0 is larger by 11 sigma. The difference is statistically significant.

What it means is that the estimates are irreconcilably different. This discrepancy first emerged in 2016 when SHOES publicized their H_0 measurement. It turned out to be different from Planck team's estimate released about two years earlier.

Since then, both SHOES and Planck teams re-calibrated their observations and improved their data analysis tools. But the error bar is getting smaller, suggesting that the discrepancy is getting worse. In light of the very low probability for these two estimates to reconcile, some astrophysicists call the Hubble constant puzzle a crisis in cosmology.[22]

6.5.3 Big Patch of Cold "Spot"

That's not all. Cosmologists have found cold "spot" on the CMB map. The temperature there is almost always below the average over an extremely vast region of the universe. As an analogy, it is as if most of the air molecules in a closed room are not uniform but concentrated at one corner. When it happens, which is very unlikely, the temperature at the big part of the room will be colder.

So it seems that the cold spot is another rebellion of Nature against the outcome expected by cosmologists. What is comprehensible about our universe is that it is mysteriously incomprehensible.

A cosmologist at the Institute of Space Science and Technology (University of Karachi, Pakistan) concludes the problem of cold "spot" in the following way.

> As it stands, our current understanding of the CMB cold spot doesn't lead us to any clear conclusions. Fully understanding it either requires much better observations, or a revision of our understanding of the universe. Hopefully, future observations from more advanced ground- and space-based telescopes can guide us towards a better explanation of this enigmatic scientific phenomenon.[23]

22. Panek, "Cosmic," para. 19–20.
23. Rahman, "Enduring," last para.

6.5.4 Non-Flat Curvature

In 2020, a paper published in *Nature* suggests that the overall statistical interpretation of the latest CMB data favors a universe with non-zero curvature. But according to the standard Big Bang model, the curvature should be zero. The universe should be flat.

This non-zero finding is yet another tension between the curvature measurement with the flatness assumption of Big Bang. Nevertheless, the authors of this peer-reviewed paper were cautious to say that it might be a statistical fluke, after stating that "the assumption of a flat universe could therefore mask a cosmological crisis where disparate observed properties of the Universe appear to be mutually inconsistent."[24]

Crisis is a strong word. What it means is that almost every piece of knowledge about the Big Bang could be wrong.

6.5.5 Cosmic Axis of Evil

And then you have this should-not-happen cosmic axis of evil in the CMB. The standard model of Big Bang assumes that at the large scale, the universe has to be isotropic—no particular direction in the sky should be favored. Also, the standard Big Bang model assumes that the distribution of mass and energy is homogeneous throughout space and over time.

But in 2005, two astrophysicists noticed a curious pattern in the map of CMB generated by NASA's WMAP satellite. They found that against their "prediction," some hot and cold spots in the CMB are not distributed randomly. As shown in Figure 6.2, they are aligned with an axis. It is dubbed the axis of evil.[25]

The axis is "evil" because it opposes the sacred cosmic Copernican principle—the orthodox foundation of big cosmic evolution. We have been told that our planet is insignificant in the cosmos. Sun is just an ordinary star at the outskirt of the Milky Way, which is just another spiral galaxy in the vast universe with no preferred direction. That is the gist of this cosmic Copernican principle.

24. Di Valentino et al., "Planck," abstract.
25. Land and Magueijo, "Examination," para.2.

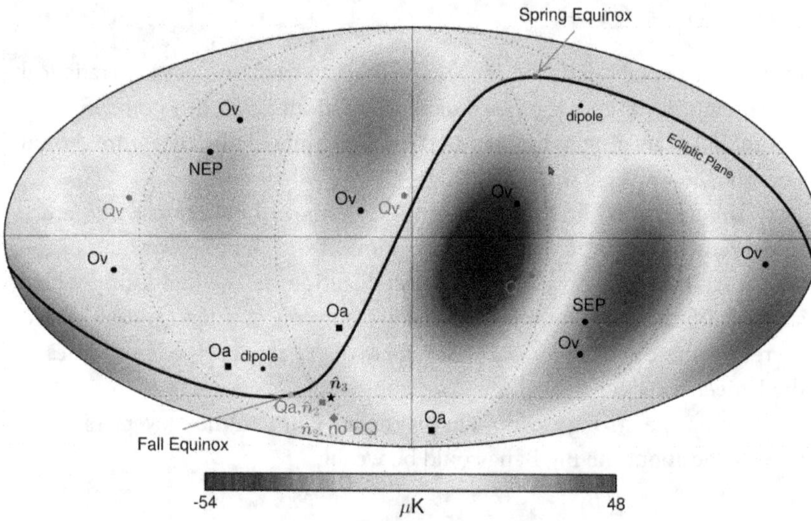

Figure 6.2. The S-shape curve is the axis of evil. It is correlated with the ecliptic plane of our planet. The spring and fall equinoxes are located on the axis. The SEP at the lower right portion is the South Ecliptic Pole. The NEP at the upper left portion is the North Ecliptic Pole.

Shaking this foundation of Big Bang is as evil as it is to deny that Jesus had resurrected. This evil axis, first found through WMAP data and then later confirmed by the more precise Planck data,[26] challenges the very core of Big Bang.

What is even more "evil" about this axis is that it seems to point toward our Earth as the center of the universe. The S-shape dark curve in Figure 6.2 has two points corresponding to the spring and autumn equinoxes. The atheist Lawrence Krauss—who told us to "forget about Jesus"—said,

> "But when you look at CMB map, you also see that the structure that is observed, is in fact, in a weird way, correlated with the plane of the earth around the sun. Is this Copernicus coming back to haunt us? That's crazy. We're looking out at the whole universe. There's no way there should be a correlation of structure with our motion of the earth around the sun—the plane of the earth around the sun—the ecliptic. That would say we are truly the center of the universe."[27]

He went on to say in his conversation with Edge.org on July 5, 2006.

26. Aron, "Planck," para. 7.
27. Krauss, "Energy," second last para.

> The new results are either telling us that all of science is wrong and we're the center of the universe, or maybe the data is simply incorrect, or maybe it's telling us there's something weird about the microwave background results and that maybe, maybe there's something wrong with our theories on the larger scales.[28]

In the context of his conversation, "all science" is a hyperbole for all the areas of historical science related to the models of evolutionary universe, in particular, the standard Big Bang model. Operational science, on the other hand, has no "evil." To his credit, Krauss didn't beat around the bush when he saw the axis of evil.

In the history of big evolution, CMB is said to be the cosmic fossil more than thirteen billion years old. The solar system and Earth are said to be formed much later, about five billion years ago according to many big evolutionists. So, the solar system wasn't even there when the CMB was formed. Why should the elliptical plane of the Earth be aligned with the cosmic axis of evil?

The Big Bang model was developed with the assumption that the location of our planet isn't special. But the CMB provides the fact that the Earth's location is special. Do you see a contradiction?

In math, a method of proof is called proof by contradiction. We start with an assumption, go through many logical steps, and come to the conclusion that the result contradicts the assumption. So the claim with that assumption must be wrong. In this sense, Big Bang is a wrong theory on the true natural history.

6.5.6 Vacuum Catastrophe

The biggest problem many mainstream cosmologists wouldn't dispute is the contradiction between the prediction by the Big Bang model and the measured value of the cosmological constant. It plays the role of vacuum energy density in the standard ΛCDM model.

The cosmological constant is many orders of magnitude smaller than the theoretical prediction. By how much? The observed value of Λ in the Big Bang paradigm is over 120 orders of magnitude smaller than the density predicted by the quantum field theory.

In other words, the discrepancy between theory and observation is an incredible 120 orders of magnitude.[29] Physicists all agree:

28. Krauss, "Energy," last para.
29. Adler et al., "Vacuum," 620.

> This is probably the worst theoretical prediction in the history of physics! Nobody knows how to make sense of this result.[30]

To appreciate this difference, it is as if a boy one-meter tall is compared to the diameter of the universe. Even this illustration of a boy versus the expanse of the universe is severely understated. In fact, the unimaginably big discrepancy is not just one universe but the swathe of one googol universes chained together. A googol, 10^{100}, is a large number, written as one followed by one hundred zeros.

Have you heard of this severe problem? What we usually hear is how impressive the Big Bang model fits the data. Most mainstream cosmologists sweep this problem under the carpet out of sight and continue their observations, hoping that one day the cosmological constant problem may go away. Some cosmologists continue to teach as if the Big Bang is the hard and proven fact beyond doubt.

Understandably, cosmologists seldom mention this catastrophe. It may jeopardize their research grant and journal publication. Similar to the case of renormalization in perturbative quantum field theory, many hope that this problem may be eventually solved—with more research funding.

But given such a big discrepancy that lies at the very core of the model, we have reason to doubt the picture painted by mainstream cosmologists.

Is Big Bang really the origin of the universe?

6.6 Is there a "New Genesis"?

My not-so-grand theory of fractional quantum Hall effect doesn't suffer from such puzzles, tensions, and crises. In fact, the word "crisis" isn't in the vocabulary of operational science. Why? It is real science that can be experimentally tested—repeatedly—on Earth by anybody in principle and in practice.

The experiments can be conducted today or a few decades later. They will always give you the same fractions when the experimental conditions are the same. This set of conditions is the *cause* of the fractional quantum Hall *effect*. Operational science is about repeatable cause to effect.

By contrast, historical scientists guesstimate the *unobservable* historic causes with the observable effects. These non-unique historic causes are the theories. They are analogous to the multiple suspects of a cold-case homicide. Since there is only one universe, only one of the many speculations can be true. But it is possible that all the speculations can be wrong.

30. Hobson et al., *General*, 187.

Evolutionary cosmology according to the Friedmann equations is a version of history by extrapolation back into dark deep time. It is a philosophical proposition. Hard facts inconsistent with the standard Big Bang model are no longer ignorable. The lithium problem, the Hubble constant tension, the vast expanse of cold spot, the possibly non-zero curvature, and the axis of evil, and so on, are facts. We can't dismiss all of them as statistical illusions anymore.

We have an Antarctic blue whale in the room. The discrepancy between the empirical value of the cosmological constant and the theory is of the catastrophic googol order of magnitude. Big Bang cosmology is an edifice built on sand of one crisis after another. Should we blindly believe what the textbooks are telling us about the dubious and probably false genesis of the cosmos?

As Augustine said, "We ought not to believe anything unadvisedly on a dubious point."[31] Big Bang has various dubious points.

Is there a "New Genesis"?

31. Drake, *Discoveries*, 175–176.

7. What are the Big Problems of Inflationary Big Bang?

> We don't have a strong test of what happened earlier in time. We have theories, but not tested. Theories of course, any bright physicist can make up theories. They could have nothing to do with reality.[1]
>
> —JAMES PEEBLES

> . . . for in the sciences the authority of thousands of opinions is not worth as much as one tiny spark of reason in an individual man.[2]
>
> —GALILEO

7.1 Presuppositions of Historical Science

As DISCUSSED EARLIER, IN modern cosmology, Big Bang is a model of extrapolation by faith. Along the journey dictated by Friedmann's calculus back to the dark deep past, big evolutionists must believe that not a single accident or surprise would occur. Nature, however, is full of surprises.

Thus, modern cosmology assumes that the universe began from an extremely small thing. Can you imagine the whole Earth is squeezed into a ping pong ball? It's not just the Earth. Cosmologists are talking about the entire universe. They compress it into a cosmic egg googolplex times smaller than a pea. They don't tell you exactly how small it really *was*, because no one knows how to measure it.

1. Couronne and Ahmed, "Top," para. 11–13.
2. Drake, *Discoveries*, 134.

Big Bang is a possible version of history. Now, history isn't observable. To prove this point, we just need to ask, "Do you have a photograph of the moment of Big Bang?" The answer is obvious. So, to make up for the lack of direct observation, big evolutionists need some philosophical presuppositions for this historical science.

One of them is uniformity.[3] It is a worldview that the same natural laws operating today had always operated in the past. At first glance, uniformity seems intuitive and reasonable. But have we tested the natural laws indeed operated like today dark deep time ago?

Uniformity is an assumption of historical geology—the present is the key to the past. James Hutton, widely considered to be the father of historical geology, imposed uniformity as follows. The past *history* of our globe *must* be explained by what can be seen to be happening now.[4]

The laws of Nature describe what *is* happening now around us. Historical scientists must believe that they don't change since the dawn of history. But where did these invariant laws come from? C.S. Lewis, a former atheist, succinctly answered the question this way:

> Men became scientific because they expected Law in Nature, and they expected Law in Nature because they believed in a Law Giver. In most modern scientists this belief has died: it will be interesting to see how long their confidence in uniformity survives it.[5]

As Kepler demonstrated, scientific research is a form of adoration of the Law Giver through faith in Christ. For Christians, the motivation for scientific research is to glorify God by unveiling the wonders *in* the universe He created.

Most modern scientists, on the other hand, while they expect the laws of Nature, they nonetheless deny categorically any Law Giver. Instead, they bet their money on uniformity because they need it to answer big questions about the origins. Some of them need it to do historical science.

According to geochronology, the Hadean eon is supposedly the earliest Earth's history. If we start the calendar running from the first day of the Hadean eon, we are supposed to be now in Year 4 billion 500 million. Under uniformity, you can't arbitrarily set the calendar otherwise.

3. Another name is uniformitarianism. I follow C. S. Lewis's choice of word.

4. National Park Service Staff, "Geologic," para. 1.

5. Lewis, *Miracles*, 140.

Therefore, an extra dimension of historical chronology must be added to any big evolution model. Geologists presuppose uniformity so as to extrapolate from the known present age to the unknown Hadean eon.

But is uniformity really true? If this philosophical assumption turns out to be invalid, all who are trying to construct the history of everything from its origin may have to change their worldview quite a bit.

Scientists are humans. Like everyone, they will be unhappy if their professional fields are critiqued. Scientists become very unhappy when their turfs are facing the perceived threat of being marginalized. As a result of the criticisms, the last thing any scientist wants to see is a cut in research funding.

Therefore, to defend materialism that encompasses uniformity, big evolutionists can't afford to allow a "Divine Foot in the door." Richard Lewontin, a renowned figure of big evolution, wrote,

> Our willingness to accept scientific claims that are against common sense is the key to an understanding of the real struggle between science and the supernatural. We take the side of science in spite of the patent absurdity of some of its constructs, in spite of its failure to fulfill many of its extravagant promises of health and life, in spite of the tolerance of the scientific community for unsubstantiated just-so stories, because we have a prior commitment, a commitment to materialism. It is not that the methods and institutions of science somehow compel us to accept a material explanation of the phenomenal world, but, on the contrary, that we are forced by our a priori adherence to material causes to create an apparatus of investigation and a set of concepts that produce material explanations, no matter how counter-intuitive, no matter how mystifying to the uninitiated. Moreover, that materialism is absolute, for we cannot allow a Divine Foot in the door.[6]

Apparently, Lewontin was willing to accept "scientific claims that are against common sense" and "patent absurdity of some of its constructs." He also honestly admitted "the tolerance of the scientific community for unsubstantiated just-so stories." He observed that some "scientific" claims of big evolution are "nonsensical, absurd, unsubstantiated," and "mystifying to the uninitiated."

Extrapolation and storytelling are pretty much what historians are doing when they construct a history from ancient manuscripts and archaeological artifacts. These precious records correspond to observation

6. Lewontin, "Billions," 31.

data of natural science. The job of a historian is to infer the most likely cause from the effects. These effects were brought about by the *true* historical cause, which is unobservable.

In the same way, given the observation data, which are the effects, big evolutionists propose various models of origin that are consistent with the effects. But is one of these models the *true* origin in history? Logically, it can be one of them. It can also be none of them.

Peebles, the 2019 Nobel laureate, says, "Theories of course, any bright physicist can make up theories. They could have nothing to do with reality."[7] They could have nothing to do with the true origin of the universe.

As an example of an attempt to construct natural history with uniformity, a story about the origin of our planet goes like this.

> . . . five billion years ago, it did not exist. Where the Earth now sits in the outer reaches of the Milky Way galaxy, there was just a large cloud of gas and dust. This is the astonishing story of what transformed that cloud dust into the planet we know today . . . It all begins with a big cloud of gas and dust, little particles of dirty stuff, grains of sand, grains of silicon.[8]

This story is interspersed with present-day observations such as the so-called molecular clouds in space. They make the story look like science. Molecular clouds are facts by observations. But to say that they are the "debris of hundreds of dead stars billions of years ago" is an *interpretation* under the presupposition of uniformity. Of course, dead stars are no longer observable. Are they? Historical science can't proceed without introducing unobservable and unknow stuff.

No scientist has ever observed the *process* of how our Earth was formed, which is said to take a very long time. Present-day observations are then added to provide a scientific spin to the story. But how can anyone be so sure that the story is really the true history of the Earth?

Nobody was there to witness the formation of our Earth and the processes of big bio evolution. We can observe a process that lasts five hours. But nobody has ever experienced five billion years.

7.2 Problems of Zero and Planck's Time

In the same vein, Big Bang cosmology is a story about the past. It has the resemblance of physics, as it borrows the repeatable operational science of

7. Couronne and Ahmed, "Top," para. 11–13.

8. Naked Science, "Birth."

particle physics to tell a story shortly after the Big Bang. It is said to occur at the unmeasurable Planck time billions of years ago. Cosmologists postulate that at that moment, the universe was almost vanishingly small and almost infinitely dense. How small is small, and how dense is dense, no one can pin down a scientifically verifiable number.

All that can be said is that if the Friedmann's scale factor becomes zero, there will be a singularity at which all the laws of physics fall apart. Namely, uniformity fails. So, there is a big problem when the scale factor is exactly zero. This is also because in the process of solving the Friedmann equations, we need to multiply both sides of one of these equations with the scale factor. In math, such multiplication of both sides of an equation is valid only if the multiplier is strictly non-zero. Otherwise, it can lead to "1 = 2," which is a contradiction.

So cosmologists can't make the scale factor equal to zero exactly. Moreover, at time 0, the universe was just a point and the density was infinite. Infinity is not a number. Topologically, a point has no dimension. Cosmologists don't have a compelling story to turn the zero-dimensional point into our empirically observed three-dimensional space which we experience every day. No historical science can change topology.

To avoid this problem, big evolutionists first pledge their faith and trust in uniformity. They retrace the deterministic trajectory governed by the Friedmann solution all the way from today back to billions of years in the past. They stop the trajectory at the so-called Planck time before the purported beginning of the universe. It is a really tiny fraction of time from time zero—the beginning.

The Planck time is a concept of quantum mechanics. How long is the Planck time? It is about 5.4×10^{-44} second. When rounded up to 10^{-43} second, this mind-boggling amount of time is one after forty two zeros to the right of the decimal point:

0.0000000000000000000000000000000000000000001.

Over this time span, photon, the fastest runner of the universe, appears to stand still. It barely covers 10^{-25} nanometer.

Obviously, the current technology is incapable of measuring the passage of time at this ultimate granularity in the lab. Even if universe could be created experimentally, it would be impossible to measure the moment of Big Bang.

Sir Roger Penrose, who won the 2020 Nobel prize in physics for his theoretical research on black holes, put forth a thought-provoking argument concerning the measurement of time. To measure time, you need a clock. The precision of a clock ultimately comes from the fact that it has mass. A sketch of the reason he gave is as follows.

Penrose highlighted two equations of the twentieth century physics, $E = mc^2$ of Einstein's theory of special relativity, and $E = h\nu$ of quantum mechanics. Here the symbol h is the Planck constant and ν is the wave frequency of a quantum particle. Frequency is the reciprocal of time. These two simple yet fundamental equations of physics lead Penrose to the conclusion that, up to a multiplier (c^2/h), mass and frequency are equivalent. Specifically, the precision of a clock is dependent on the frequency. This is essentially how an atomic clock works.

Therefore, through this relation, $\nu = (c^2/h)m$, we learn that the precision of a clock depends on mass m. Also, any particle with mass m is a clock with a precisely defined frequency ν.

Penrose went on to argue that if we didn't have mass, then we wouldn't have a clock as the frequency becomes zero when the mass is zero. If we don't have a clock, we don't have a measure of time. Noting that a meter in the SI unit is now defined as the fraction of a second that light travels in the vacuum, so what defines the length is time. Without time, we don't have a measure of distance either. It follows that the ruler or scale to measure length requires mass too.

If there is no mass, the scale becomes irrelevant; big and small are equivalent. We have the so-called conformal invariance. This invariance is a starting point for Penrose to propose his conformal cyclic cosmology. Penrose wants to know what came *before* Big Bang.[9]

After some excitement in a small community of scientists, it becomes clear that Penrose's theory isn't supported by the data of microwave background.[10] This version of cosmic history isn't accepted as mainstream because it doesn't tally with observations.

In any case, let's time-travel back with the ticket of uniformity to the moment of the supposed Big Bang. We would find the energy so high that particles would become effectively mass-less. In the absence of mass, there is no way to measure time, and thus no scale to measure length either. The upshot is that Friedmann's solution that describes how the size of the universe changes with time won't work anymore.

The implication of Penrose's argument is that scale is meaningless without time. In other words, without a measure of time, the measure of scale becomes impossible. As a result, you can't tell time and scale when the universe was presumably in the mass-less state.

9. Penrose, "Cycles," part 3.

10. Jow and Scott, "Re-evaluating," sect. 4.

Big Bang theorists assume that the universe was initially smaller than a proton. Some theorists call it "nothing." This name is quite misleading but it is catchier than a "primordial egg" of pure energy, which is something. It will arrest more media attention to say, "Nothing created the universe." rather than "The universe started with a primordial egg."

As it turns out, "nothing"' is a grotesque misnomer. It is not really nothing but a postulated state in some sort of quantum field-theoretic framework. Depending on who you talk to, "nothing" has a different meaning. "Nothing" makes modern cosmology confusing. Big evolutionists have been trying to fabricate something from "nothing."

But no one really has a clue about the origin, as Friedmann's solution is disallowed to run its full course. Some theoreticians claim that the universe began with quantum fluctuations. Some say it is one of the baby universes. Adding to the confusion are those with the word "origin" in the official announcements, such as "Understanding the Origin of the Universe."[11] One thing is clear, though. All the models are philosophical speculations about the true origin.

7.3 Hawking's Fallacies

Yet many atheists continue to chant the mantra that our universe spontaneously began all by itself is science. Hawking said that nothing created the Big Bang. The origin of our universe is nothing. I have great respect for Hawking as a brilliant scientist. He is also a brilliant best-seller author. Whenever I use his name in the critique below, I am criticizing his ideas and arguments only. This is the same rule I have been practicing for other people.

7.3.1 Hawking's Pangu

To understand Hawking's writing in chapter 1 of his posthumous book, *Brief Answers to Big Questions*, we need to first unpack what he meant by "nothing." The pre-existing universe has zero energy is probably what he meant. Hawking thought that "the universe was spontaneously created out of nothing, according to the laws of science."[12] Positive and negative energies "add up to zero."[13] It seems that Hawking's nothing refers to the state of no energy.

11. Ushijima, "Understanding."
12. Hawking, *Answers*, chapter 1, para. 8.
13. Hawking, *Answers*, chapter 1, para. 18; para. 22.

Hawking claimed that no energy is needed for our universe to create itself. He argued that from his nothing of zero energy comes something "*E* + (-*E*)." These two terms "add up to zero."

He drew on an analogy as follows.

> Imagine a man wants to build a hill on a flat piece of land. The hill will represent the universe. To make this hill he digs a hole in the ground and uses that soil to dig his hill. But of course he's not just making a hill—he's also making a hole, in effect a negative version of the hill. The stuff that was in the hole has now become the hill, so it all perfectly balances out. This is the principle behind what happened at the beginning of the universe.[14]

The symbol *E* represents the matter and energy *in* the universe. It's the hill in this analogy. The hole is the analog of the negative energy -*E*. "The stuff that was in the hole has now become the hill, so it all perfectly balances out." It's a nice analogy.

But how does the action of digging happen? Where did the "flat piece of land" come from? The man who wants to build a hill, who is he? Where did he get the energy to dig?

Hawking didn't really explain the origin of the universe. He just kicked the can down the road of unknown "flat piece of land."

To dig a hole, the man needs a shovel. Hawking's analogy is rather similar to the Chinese myth of Pangu creation. I read this fairy tale when I was a boy. At the beginning, there was nothing, a chaos. Pangu the giant used an axe to create the world by splitting the chaos.[15] We easily discern that Hawking's man is Pangu, the "piece of flat land" is chaos, and the shovel is the axe. "It is a "romantic story of China."[16]

But biblical creation is totally different. In the beginning was the Word. So time started "in the beginning." At the same instance, God created the canvas of the universe (heavens) and Earth. There was no Hawking's man, no Pangu. God didn't use any shovel. He didn't swing any axe.

7.3.2 Problems of Negative Energy

Now, what is -*E*, the negative energy? In physics, gravitational potential energy is negative. Hawking claimed that "space itself is a vast store of

14. Hawking, *Answers*, chapter 1, para. 17.
15. Confucius Institute for Scotland, "1," para. 2.
16. Confucius Institute for Scotland, "1," para. 3.

negative energy."[17] Thus -*E* is supposedly the gravitational potential energy stored "in space." Clearly, the sum of *E* and -*E* is equal to zero, which is Hawking's nothing. So Hawking argued that during the origination of the universe, Big Bang's something *E* from his nothing didn't violate the fundamental law of energy conservation.

Is Hawking argument valid?

I need to point out that space as a "vast store of negative energy" is a misleading oversimplification even for Hawking's readers. In physics, the negative energy is determined by the distance between two objects with mass. So two massive objects are needed. How can gravitational potential energy begin to exist if there is only his nothing of zero energy alone in the beginning?

Hawking furthermore said that the universe itself initially was very small, "perhaps smaller than a proton."[18] So now we have his nothing *and* small space. Hawking was silent on why there is something smaller than a proton to begin with. Where did this mysterious unobservable space come from? With our earlier discussion, we can see that this tiny space is the result of backward extrapolation from today's universe.

Hawking thought that his nothing could cause our universe. But he required the negative gravitational potential energy to work. But this trump card of Hawking is questionable from the physics of gravity. It is questionable because Einstein's *classical* gravity of general relativity doesn't work in his quantum space smaller than a proton.

It is assumed that the universe in the beginning was so small that only quantum mechanics could be applied. Therefore, without the laws of gravity, we won't have the potential energy of gravity in the beginning. Without this negative energy to balance the zero-energy budget of nothing, the law of energy conservation is violated. Something from Hawking's nothing goes against basic physics.

Hawking's idea of nothing "0" to something $E + (-E)$ will surely violate the law of energy conservation. The only solution to obey the law is to have $E = 0$, which is nothing. It means Hawking's nothing gives you nothing. This is the logical conclusion. Hawking's nothing can't generate anything.

Therefore, there is no such thing as the ultimate free lunch. We are back to the big question which Hawking thought he had solved it. Where did the enormous energy *E* of Big Bang come from?

How can a distinguished theoretical physicist make such a mistake in physics? In an earlier chapter, we have seen that Einstein erred in his

17. Hawking, *Answers*, chapter 1, para. 19.

18. Hawking, *Answers*, chapter 1, para. 23.

evaluation of Friedmann's solution, which shows that the universe can expand. Einstein believed that the universe was static. This belief made him jump to the wrong conclusion that Friedmann's solution couldn't possibly be correct. Moreover, to make the universe static, Einstein introduced the cosmological constant. Later, he rescinded it and described this embarrassing episode as the biggest blunder of his life.

Even Einstein, who by any estimation was greater than Hawking, made at least two mistakes. These two mistakes in Einstein's superstar career could be avoided. If he didn't hold a philosophical bias against solutions other than the static universe, he wouldn't have to reverse his comment on Friedmann's paper.

7.3.3 Hawking's River

Anyway, with a strong belief that his nothing could cause the Big Bang, Hawking discussed the implication of his philosophical claim as follows.

> So what does this mean in our quest to find out if there is a God? It means that if the universe adds up to nothing, then you don't need a God to create it. The universe is the ultimate free lunch.[19]

But who made the ultimate lunch free? Hawking answered this question by starting with a river flowing down a mountainside by gravity today. He then asked what caused the river. And what caused the cause that caused the river. Eventually, it went all the way back to a tiny space before Big Bang.

If there is no river (nothing) today, will any cosmologist have the motivation to extrapolate back in time? Hawking already had a river (something) today in the universe. So he went all the way back in time to the Big Bang. Then from the Big Bang, Hawking went forward in time with big evolution to explain the river we see today. This argument seems a bit circular.

7.3.4 No Time and No Cause

Hawking claimed that his nothing of zero energy is possible to cause the universe to exist.

> But here's the crucial bit. The laws of nature itself tell us that not only could the universe have popped into existence without any assistance, like a proton, and have required nothing in

19. Hawking, *Answers*, chapter 1, para. 21.

> terms of energy, but also that it is possible that nothing caused the Big Bang. Nothing.[20]

By philosophically insisting that the universe came from his nothing, Hawking's argument violates the fundamental law of cause and effect. This law of logic states that any effect that begins to exist has a cause. Universe began to exist. It must have a cause. But Hawking said, "it is possible that nothing caused the Big Bang. Nothing." Can no energy cause the Big Bang to produce our universe?

On time, Hawking wrote,

> The explanation lies back with the theories of Einstein, and his insights into how space and time in the universe are fundamentally intertwined. Something very wonderful happened to time at the instant of the Big Bang. Time itself began.[21]

The assertion that "time itself began" from non-existence without a cause is once again a breach of the logic of cause and effect. No law of science can tell us that time must and can begin by itself.

By contrast, the Bible proclaims that "In the beginning was the Word, the Word was with God, and the Word was God." and "In the beginning God created the heavens and the earth." God created time "in the beginning," the heavens (that's space), and the earth (that's matter). The origins of time, space, and matter are supernaturally created. God is the First Cause. He is eternal. He transcends time.

James Clerk Maxwell was a Christian theoretical physicist. He invented electromagnetism, which is the foundation of electricity. He said,

> Science is incompetent to reason upon the creation of matter itself out of nothing. We have reached the utmost limit of our thinking faculties . . . because matter cannot be eternal and self-existent it must have been created.[22]

Nevertheless, Hawking thought that he "finally found something that doesn't have a cause, because there was no time for a cause to exist in."[23] This remark is quite nonsensical as well.

If there was no time for a cause to exist in, then Hawking's nothing has no time to begin to exist in either. Otherwise, we would observe strange things when the law of cause and effect is violated.

20. Hawking, *Answers*, chapter 1, para. 24.
21. Hawking, *Answers*, chapter 1, para. 25.
22. Maxwell, "Molecules," 376.
23. Hawking, *Answers*, chapter 1, para. 28.

As a thought experiment to illustrate this point, suppose Hawking opens a credit card account with no money (nothing) deposited at the Big Bank. Suddenly, as "time begins itself" with a bang, it shows a debit of -*E* pounds. Suddenly, he owes the bank *E* pounds when the bank changes its name to Big Bang.

By Hawking's logic, he can't claim that the man who wants to build a hill has misused his credit card. Where does the money *E* go? It becomes the fund for building the one and only one university. It is fully owned and operated by the bank (the laws of science). The university is the ultimate free lunch.

By saying the Law of Nature allows the Big Bang to generate itself spontaneously without a cause, Hawking was trying to make the point that Nature is all powerful. Of course, Hawking has no empirical proof of his nothing.

But the laws of Nature are the description of how Nature operates and not how matter is created from nothing. Big Bang from nothing is as miraculous as a car without an engine moving uphill without a reason. Effectively, Hawking's nothing is a blatant violation of the law of cause and effect.

Like many atheists, Hawking wanted to avoid the supernatural Creator at all costs. In so doing, he had to resort to saying that his nothing caused the Big Bang, which is as good as claiming that Big Bang had no cause for it to begin. Even if Hawking is correct, his claim of nothing causing something to exist is a philosophical position. It is unscientific, for there is no way to test his claim.

By violating the law of cause and effect, Hawking's answer to the big question of how did the cosmos begin by nothing—i.e., with zero energy—is a grand fallacy. It contradicts the foundation of naturalism of not allowing any infringement of the laws of Nature. The seemingly profound answer self-destructs. He also had not answered the big question.

7.3.5 Free Lunch in Information?

There are still a few more problems. Theoretically, must the zero-energy configuration be $E + (-E)$? What about $0 + (-0)$? This configuration of nothing with E=0 is physically more probable. From Hawking's nothing we get nothing. This configuration of positive zero plus negative zero has no contradiction with the law of energy conservation and the law of cause and effect.

Also, is there a theory to predict the value of *E* from first principles?

Furthermore, from the viewpoint of information content, the configuration of Hawking's nothing, a piece of flat land of "o" has one bit of information. But the configuration of a hill plus a hole, $E + (-E)$, has three bits. They are E, $-E$, and the addition. The big question is, what causes the information content to increase from one bit to three bits? There is no known law of physics that can increase the information content spontaneously without an external input to the system.

7.3.6 Misunderstanding God of the Bible

Hawking held a wrong concept of God. In his mind, God can't do anything. He wrote,

> The basic assumption of science is scientific determinism. The laws of science determine the evolution of the universe, given its state at one time. These laws may, or may not, have been decreed by God, but he cannot intervene to break the laws, or they would not be laws. That leaves God with the freedom to choose the initial state of the universe, but even here it seems there may be laws. So God would have no freedom at all.[24]

Hawking believed in scientific determinism. He showed us how uniformity can be described: "The laws of science determine the evolution of the universe, given its state at one time." Hawking's "God" is not omnipotent and not sovereign. He "cannot intervene to break the laws, or they would not be laws." So, "God would have no freedom at all." Hawking put the laws of science above God.

But clearly, any Christian can tell that Hawking's "God" isn't God of the Bible. When Jesus was walking on Earth, He was bound by the law of gravity. But by His sovereign will, Jesus walked on a stormy Sea of Galilee, against gravity. Jesus wanted the disciples to know that He is the Son of God. Every deliberate miracle is a suspension of the laws of Nature at a pinpoint time and location. Other parts of the cosmos operate normally under the law of gravity when Jesus was walking on water.

Toward the end of the first chapter of his book, Hawking wrote,

> When people ask me if a God created the universe, I tell them that the question itself makes no sense. Time didn't exist before the Big Bang so there is no time for God to make the universe in.[25]

24. Hawking, *Answers*, chapter 1, para. 6.
25. Hawking, *Answers*, chapter 1, para. 29.

Usually, the "God" Hawking had in mind is not God of the Bible. He took it to mean "the embodiment of the laws of nature."[26] He used it "in an impersonal sense, like Einstein did."[27]

But in this context, Hawking was referring to God of the Bible because the question he posed is whether "a God created the universe." Hawking gave his *opinion* that God of the Bible couldn't possibly create the universe because time didn't exist before the Big Bang and so "there is no time for God to make the universe in."

Hawking's atheistic worldview absolutely must not allow any god to create. But his biased opinion has a self-defeating fallacy. We can use his argument to assert likewise that "Time didn't exist before the Big Bang so there is no time for Hawking's nothing to make the universe in" either.

Furthermore, Hawking's understanding of God of the Bible is fundamentally flawed. He thought that God of the Bible was bounded by time. But the sixty six books in the Bible tell us that God (*Elohim*) transcends time and space. God of the Bible is not limited by time, "who is and who was and who is to come, the Almighty."[28] He exists even if there is no time and no space. God of the Bible is the First Cause of everything. He is transcendent and eternal.

To wrap our mind around His infinity, either in the sense of space or time, is impossible. In Christian theology, God created space, time, and matter out of absolutely nothing, and He set the arrow of time flying in the beginning. Through His infinite wisdom and by His words, God created the laws of Nature. At His own sovereign will, God can enter the cosmos He created as Jesus did.

7.3.7 Hawking's Atheism

To conclude chapter 1 of his book, Hawking answered the question of "Is there a God?" with a resounding "No." In the mechanical tone of scientism, he wrote,

> Do I have faith? We are each free to believe what we want, and it's my view that the simplest explanation is that there is no God. No one created the universe and no one directs our fate. This leads me to a profound realisation: there is probably no heaven and afterlife either. I think belief in an afterlife is just wishful

26. Hawking, *Answers*, chapter 1, para. 5.
27. Hawking, *Answers*, chapter 1, para. 6.
28. Revelation 1:8.

> thinking. There is no reliable evidence for it, and it flies in the face of everything we know in science.[29]

Now, these are not statements of science but Hawking's personal belief. As Frank Turek, the founder of Cross Examined quips, "'Science' doesn't say anything—scientists do."[30] The desirable outcome of "no one directs our fate" allows each atheist to be god. "No heaven and afterlife" is also desirable, for then there is no Law Giver watching over us, and so no ultimate judgment. Everyone is god.

By definition, anything unilaterally done by the divine being is supernatural and thus the six-day creation is a miracle. Christians admit that creation is a miracle. But the logical conclusion from Hawking's argument is also a miracle. This is because his arguments breach the law of causality and the law of energy conservation.

Besides, Hawking is wrong about afterlife too. It is not a natural phenomenon. Granted, miracles are non-scientific. Scientific methods are not designed to study the supernatural stuff. Rather, they are intended for studying the natural world that can be observed repeatably. To apply the scientific method on entities that transcend Nature is like trying to put a square peg into a round hole.

7.3.8 The Energy-Time Uncertainty Principle

Not only that, Hawking's argument violates the energy-time uncertainty principle of quantum mechanics. At time zero, time is known; it is exact. So there is no uncertainty in time, which means that the uncertainty in energy must be infinite. So, in principle there is no way to estimate energy. But Hawking's nothing has zero energy exactly. On the other hand, if energy is known exactly to be zero, then time must be infinitely uncertain. So we can't know when did Big Bang occur. Either way, Hawking's nothing won't work.

Hawking's book was targeted at the general public. He simplified a lot of physics, and he surely was aware of this violation. To avoid it, let's consider instead the Planck time of 5.4×10^{-44} second as the uncertainty. To estimate the uncertainty in energy *E*, first, we divide the Planck constant *h* by 2 pi to obtain the so-called reduced Planck constant, which is about 1.05×10^{-34} with the unit of Joule second. We then divide the reduced Planck constant by the Planck time to obtain 2.78 billion Joule for the uncertainty in energy *E*.

29. Hawking, *Answers*, chapter 1, para. 30.
30. Turek, "Science," para. 2; *Stealing*, chapter 6.

Now, energy can be expressed as the Boltzmann constant times temperature (in Kelvin). Since we know the Boltzmann constant, we can express energy in terms of temperature. Like foreign exchange, the conversion rate is one Joule to 7.24×10^{22} Kelvin. Therefore the energy uncertainty of 2.78 billion Joules correspond to the temperature uncertainty of about 2×10^{32} Kelvin. With this simple calculation, cosmologists might have estimated that the temperature of the baby universe at the Big Bang was about this order of magnitude.

Cosmologists can give any value to E of at least 10^{32} Kelvin. Nobody can measure this outrageous temperature, and no one can deny it either. In short, E is empirically unknown.

Who can reasonably believe that the universe of unknown E popped out at time 0? Using the unknown to explain the origin of the universe is what atheists like Hawking do.

Some atheists believe the laws of Nature can create our universe. But Hawking's worldview raises a bigger question. Where and how did the laws of Nature come about?

What is Nature? Isn't Nature the universe and everything in it? If Nature did not yet exist, could there be laws of Nature to govern it? Using Hawking's argument, time didn't exist before the Big Bang so there is no time for the laws of Nature—Hawking's god—to make the universe in. Nothing seems to make sense.

7.4 The Paradigm of Inflation

To solve three foundationally serious problems of the extant Big Bang model, Alan Guth proposed to add an ingredient called "inflation." This idea in 1981 is out of the box. The inventor of the inflation paradigm wrote,

> Although it is called the "big bang theory," it is not really the theory of a bang at all. It is only the theory of the aftermath of a bang. It elegantly describes how the early universe expanded and cooled, and how matter clumped to form galaxies and stars. But the theory says nothing about the underlying physics of the primordial bang. It gives not even a clue about what banged, what caused it to bang, or what happened before it banged. The inflationary universe theory, on the other hand, is a description of the bang itself, and provides plausible answers to these questions and more. Inflation does not do away with the big bang

> theory, but instead adds a brief prehistory that joins smoothly to the traditional description.[31]

Specifically, at 10^{-34} second, a runaway process called "inflation" is said to happen. Suddenly, the size of the universe ballooned by a factor of 10^{40}. This inflation is faster than the speed of light.

Of course, this idea can't be directly tested. Guth justified it by showing that his model could explain certain observations collected by astronomers. Since then, many research papers based on the paradigm of inflation have been published.

If you surf the internet, you will surely come across online textbooks or blogs that provide the timeline of the universe. You will find inflation in the timeline as if it is a scientific fact. One thing you can be sure is that no author or blogger has ever observed the cosmic inflation. For that matter, no scientist has ever observed the explosive inflation simply because it is impossible to observe. This is the danger of believing whatever the textbooks say about the history of Nature without asking questions. We may ask,

- How can you measure 10^{-34} second?
- Without observing the inflation when the universe was supposedly 10^{-34} second old, how do you know that it really had happened?

These two questions are fair challenges. The reason is that cosmologists must presuppose uniformity when they rewind the clock. The motivation for these two questions also stems from a puzzle that cosmologists seem to ignore.

To present this puzzle, let us first estimate the temperature of the universe at the beginning with the Big Bang paradigm. Astrophysicists tell us that the critical density of the universe is of the order of 10^{-26} kg per cubic meters.[32] This density depends on the Hubble constant estimated today. They also tells us that the volume of the observable universe is approximately 10^{80} cubic meters.[33] Therefore, we can multiply these two values together to obtain the content of the universe. In kilograms, it is of the order of 10^{54}. This amount of mass, by the famous $E = mc^2$ formula, is equivalent to an energy level of roughly 10^{94} Kelvin. Since energy-matter is always conserved, in the beginning, this gross amount of energy should be present. That is, the supposed Big Bang's temperature before Guth's inflation was of the order of 10^{94} Kelvin.

31. Guth, "Inflation," 30.
32. Gunn, "Dense," para. 6.
33. Deibert, "How," para. 6.

By contrast, the temperature at which inflation is said to kick in is *only* of the order of 10^{32} Kelvin.[34] Given that the microwave background is about 3 Kelvin, it took the universe about fourteen billion years to cool by thirty two orders of magnitude. This amount is minuscule compared to the super-cooling from time zero to the inflation epoch a tiny second later. How did the temperature super-cool within such an ultra-short span of time by sixty two orders of magnitude?

The other problem of the inflation paradigm is that there are many different models. Cosmologists aren't sure which is the best one. Some inflation models can produce a multitude of patches of many baby universes. Each baby is said to develop into a different kind of universe than the one we are in. In other words, our universe by accident happens to be the lucky one that can support life.

It seems that our universe is finely tuned for us to thrive on Earth. Big evolutionists always rely on randomness, accidents, and collisions in explaining the origins of Moon, Earth, and the other planets in our solar system. They can't entertain the notion of a super intelligent designer who designed an orderly cosmos fine-tuned for us.

This is one of the possible reasons why the idea of multiverse is put forth. The argument is that if there are a lot of universes, then our universe happens to be the lucky one that allows big evolution to take place. So fine-tuning is no longer an issue.

But does multiverse really explain fine-tuning? As will be elaborated shortly, the answer is an emphatic no.

Even if multiverse is true, how is it possible to verify the claim scientifically? Sabine Hossenfelder, an atheist, theoretical physicist, youtuber, and an author, concludes with the following words.

> I put together these videos because I am frustrated that scientists discard the issue unthinkingly. This is not a polemical argument and it's not meant as an insult. But believing in the multiverse is logically equivalent to believing in god, therefore it's religion, not science.[35]

Hossenfelder dislikes multiverse so much that she relegates it from science to religion. For some atheists, religion is the best way to express the lowest level of reason.

34. Chow, *Gravity*, 204.

35. Hossenfelder, "Why," para. 2.

7.5 Deflating the Inflation Paradigm

Hossenfelder isn't the only one. John Horgan, in December 2014, posted a blog in Scientific American. In it he reported his interview with an apostate of inflation. His name is Paul Steinhardt, the Albert Einstein Professor in Science at Princeton University.

During the interview, Steinhardt expressed concern about the idea of multiverse. He said,

> My concern was that the multiverse is a 'theory of anything', a proposal that allows all possible cosmological outcomes (smooth or not smooth, curved or flat, etc.) and, consequently, is not subject to empirical tests. Some claim that superstring theory allows exponentially many (or perhaps infinitely many) possibilities for the fundamental laws (masses of particles, types of forces, etc.) and that there is no guiding principle to determine which set of physical laws is more probable. The sets of laws comprise what is called the "string landscape."
>
> Combine the inflationary multiverse with the string landscape, and now one has a 'supertheory of anything': both the cosmological properties and the microphysical properties of the universe are accidental and unpredictable.[36]

Indeed, Steinhardt is one of the early researchers of the inflation paradigm. Guth's original model of inflation has some problems. Steinhdart and Linde found a way to overcome them. In May 1984, Steinhdart and Guth published a joint article in Scientific American with the title "The Inflationary Universe."

In 2002, Guth, Linde, and Steinhdart won the ICTP Dirac Medal, "for the development of the concept of inflation in cosmology." The award press release also says, "Although the history of the very early universe has not been firmly established, the idea of inflation has already had notable observational successes, and it has become the paradigm for fundamental studies in cosmology."[37] Take note of the assessment: "the history of the very early universe has not been firmly established."

About ten years after winning the ICTP Dirac medal, Steinhardt changed his mind. In June 2013, he and his two co-authors published a paper in a peer-reviewed journal. The trio argued that the data captured by the Planck satellite constrain the various models of inflation to the class that have a plateau in the inflationary quantum field. They identified three

36. Hogan, "Physicist," para. 24–25.

37. ICTP Staff, "Dirac," para. 1.

new problems of the inflation paradigm. One of them is what they call the inflationary unlikeliness problem. They say,

> All inflationary potentials are not created equal. The odd situation after Planck2013 is that inflation is only favored for a special class of models that is exponentially unlikely according to the inner logic of the inflationary paradigm itself . . . the fact that only plateau-like models are favored is paradoxical because inflation requires more tuning, occurs for a narrower range of parameters, and produces exponentially less plateau-like inflation than the now-disfavored models with power-law potentials. This is what we refer to as the inflationary "unlikeliness problem" . . . an alarm warning us that a paradigm can fail even though observations favor a class of models if the paradigm predicts the class of models is unlikely.[38]

They concluded that "recent experimental data disfavors all the best-motivated inflationary scenarios and introduces new, serious difficulties that cut to the core of the inflationary paradigm."[39]

To make their technical paper understandable to general public, Steinhardt and his two collaborators published an article in Scientific American. They also urged their fellow cosmologists to reassess the inflation paradigm. New ideas are needed to figure out how the universe actually began. Their reason is that "we should not just accept the assumption that inflation happened, especially because it does not offer a simple explanation of the observed features of the universe."[40]

7.5.1 Cosmic Inflation Debate

It turns out that that article stirred up much more than a storm in a teacup. At stake is the mainstream model of the very early universe. The magazine Physics Today ran an article published on June 5, 2017. The article reports that "the trio's aggressive reappraisal of a scientific consensus inspired an energetic rebuttal . . . from thirty three prominent physicists."[41] They include Allan Guth, Andrei Linde, Sean Carroll, Stephen Hawking, Lawrence Krauss, Martin Rees, Leonard Susskind, Alexander Vilenkin, Steven

38. Ijjas et al., "Inflationary," 263.
39. Ijjas et al., "Inflationary," 261.
40. Ijjas et al., "Pop," 38.
41. Corneliussen, "Cosmic," para. 1.

Weinberg, Frank Wilczek, and Edward Witten.[42] The last three of them are two Nobel laureates and a Fields medalist who applied theoretical physics to solve pure math problems.

The "energetic rebuttal" was published in Scientific American on May 10, 2017.[43] The rebuttal appears to perform a damage control of the damages inflicted by the trio. It goes into considerable length to describe how a class of inflationary models does a great job in explaining the data. Understandably, the rebuttal letter doesn't mention at all the foundational mismatch between observations and theories.

The rebuttal letter opens with a logically fallacious *argumentum ad populum*. ". . . there are now more than 14,000 papers in the scientific literature, written by over 9,000 distinct scientists, that use the word "inflation" or "inflationary" in their titles or abstracts . . . [the trio] are dismissing the research of not only all the authors of this letter but also that of a substantial contingent of the scientific community."[44] This argument based on the *current* popularity of the inflation paradigm is irrelevant as it doesn't tell us whether the points raised by the trio are unsound.

Let's revisit the case of Galileo. The popular view of university professors and scholars at his time was the geocentric paradigm of Aristotle and Ptolemy. Today, we know that Ptolemy's theory is deficient. It is one Galileo against many. Here, we see a parallel. Thirty three authorities of science rebuked three colleagues for challenging the consensus. Ten against one.

To the credit of both camps, it is not an attack on personality or character. But their sharp differences of opinion concerning a paradigm of historical science spilled over to the media.

In conclusion, the rebuttal letter signed by thirty three luminaries exclaimed, "Empirical science is alive and well!"[45] Yes, it is well with operational science that works repeatably. As explained in the previous chapters, we have reasons to doubt whether big evolution is empirically alive and well.

The rebuttal is highly atypical. Fifty to seventy years ago, there was also a fierce competition between the steady-state model and the Big Bang model. But the heated debate then didn't disintegrate into something that requires thirty three luminaries of science to sign a letter against the dissenting trio.

To the credit of Scientific American, the editor DiChristina included the response to the rebuttal from the trio. After paying "great respect for

42. DiChristina, "Cosmic," para. 14.
43. DiChristina, "Cosmic," para. 1.
44. DiChristina, "Cosmic," para. 3.
45. DiChristina, "Cosmic," para. 13.

the scientists who signed the rebuttal," the trio goes on to say that their distinguished colleagues have missed their key point: "the differences between the inflationary theory once thought to be possible and the theory as understood today. The claim that inflation has been confirmed refers to the outdated theory before we understood its fundamental problems."[46]

As one of the researchers during the early days of the inflation paradigm, Steinhardt of the trio has the first-hand working experience of what had happened with regard to the theoretical work on inflation. The trio wrote in their response:

> Historically, the thinking about inflation was based on a series of misunderstandings. It was not understood that the outcome of inflation is highly sensitive to initial conditions. And it was not understood that inflation generically leads to eternal inflation and, consequently, a multiverse—an infinite diversity of outcomes. Papers claiming that inflation predicts this or that ignore these problems.[47]

The trio has also set up a website to provide more details of their response to the rebuttal letter. In an FAQ they wrote, "Originally scientists thought that the outcome of inflation (a smooth, flat universe with a certain spectrum of density fluctuations and gravitational waves) was generic. Now we know it is not."[48] The central message of the trio is that the inflation paradigm produces no empirically testable prediction.

As a matter of fact, Guth himself realizes this undesirable outcome. In a peer-reviewed journal in 2007, he published a paper in which he wrote,

> In an eternally inflating universe, anything that can happen will happen; in fact, it will happen an infinite number of times. Thus, the question of what is possible becomes trivial—anything is possible, unless it violates some absolute conservation law.[49]

Under the inflation paradigm, to say that "anything that can happen will happen an infinite number of times" is quite shocking for any scientist working in operational science. It means that the inflation models have nothing to tell us. It also means that you can take anything you want as "scientifically" true. Few scientist will recognize "anything is possible" as science.

In conclusion, Guth wrote,

46. DiChristina, "Cosmic," para. 15.
47. DiChristina, "Cosmic," para. 16.
48. Ijjas et al., "Pop," FAQ.
49. Guth, "Eternal," sect. 4.

> Essentially all inflationary models are eternal . . . A crucial issue in our understanding of fundamental physics is the selection of the vacuum, which according to current ideas in string theory could be any one of a colossal number of possibilities. Eternal inflation offers at least a hope that a small set of vacua might be strongly favoured. But so far it is only wishful thinking to suppose that eternal inflation will allow us to determine the vacuum in which we should expect to find ourselves.[50]

7.5.2 Lessons from the Saga

What can we take away from this saga?

One, the earliest models of ICTP Dirac medalists were inappropriately developed. The implications physicists drew from the models are therefore a series of misunderstandings. After forty years, the picture from the inflation paradigms is that "essentially all inflationary models are eternal." It means that cosmic inflation is still ongoing even when you are reading this part of the book. The unstoppable inflation will lead to an infinite number of possible universes. To find ourselves in this fine-tuned universe is "wishful thinking."

Two, the inflationary Big Bang has become an orthodoxy. Steinhardt and his co-authors challenged the empirical testability of the inflation paradigm. We witness a battle between the orthodox and Steinhardt's newly proposed model of a bouncing universe. Being very different, only one of the two competing paradigms could possibly be correct. Even so, logically, both paradigms can be wrong.

Three, this saga also shows that these models of the very early universe can't be differentiated by experiments in the lab. Otherwise, we should be able to know which model is the right one.

By contrast, in operational science such as microbiology, we can perform experiments to decide which of the theories is correct. For centuries, naturalists believed that a variety of organisms could pop into existence spontaneously. It is similar to Hawking's universe from nothing. The alternative theory was that organisms are produced only by their parents.

In the nineteenth century, the French scientist Louis Pasteur conducted experiments to determine which of the two theories was correct. His experiments conclusively showed that microbes can't possibly arise from non-living materials. The experiments put *the* end to the debate between

50. Guth, "Eternal," sect. 7.

spontaneous generation and parental reproduction. Pasteur, a creationist Christian, concluded as follows.

> Never will the doctrine of spontaneous generation recover from the mortal blow of this simple experiment. There is no known circumstance in which it can be confirmed that microscopic beings came into the world without germs, without parents similar to themselves.[51]

If you perform the same experiments of Pasteur in our twenty first century, you will surely arrive at the same conclusion. Day-to-day operational science is age-less, alive, and well.

On the contrary, between the big inflation and the big bounce, no experiment can be performed to settle the debate between these two versions of history of dark deep time. Both of them can't be tested. Why? Both are not dealing with the subject of the present but of the past, once upon a time in the history. Experiments similar to the ones of Pasteur are not possible.

Pasteur, arguably the father of modern medicine, was a devout Christian. He stated his opposition to the materialistic explanation of the origin of things in these words.

> Posterity will one day laugh at the foolishness of the modern materialistic philosophers. The more I study nature, the more I stand amazed at the works of the Creator. I pray while I am engaged in my work in the laboratory.[52]

We can sense the spiritual joy Pasteur was experiencing as he stood in amazement at the works of the Creator he discovered. Pasteur could truly appreciate the fact that "the heavens are telling of the glory of God, and their expanse is declaring the work of His hands."[53]

7.5.3 Multiverse and Fine-Tuning

The inflation paradigm gives rise to an explosion of infinite number of universes. Some cosmologists are very enthusiastic about the multiverse. They thought this eternally reproducing monster may provide the solution for the fine-tuning problem.

But Philip Goff, a philosopher, argues that the multiverse can't possibly solve the fine-turning problem, Specifically, using the multiverse to explain

51. Gillen and Sherwin, "Louis," 43–52.
52. Literary Digest Writer, "Darwinism," 490.
53. Psalm 19:1.

fine-tuning is logically fallacious because such usage is an inverse gambler's fallacy. Goff describes this fallacy as follows.

> A visitor walks into a casino and the first thing she sees is someone rolling a double six. She thinks "Wow, that person must've been playing for a longtime, as it's unlikely they'd have such good luck just from one roll." This is fallacious for the same reason. The casino-visitor has only observed one roll of the dice, and the odds of that one roll coming good is the same as any other roll: 1/36. How long the player has been rolling prior to this moment has no bearing on the odds of the one roll the visitor observed being a double six[54]

The "longtime" here refers to a large number of rolls before observing a double six.

Every roll of a pair of fair dice is independent of the previous roll and the next roll. The probability of each roll doesn't change. This is a basic fact in the theory of probability.

We know that our universe is fine-tuned. It is one roll of double six. Getting a double six doesn't necessarily follow that it takes many rolls for it to happen. It is still extremely improbable for our fine-tuned universe to pop out.

Therefore, it is fallacious to reason like this: "If there are enough universes, with enough variation in the numbers in their physics, then it becomes statistically likely that one will happen to have the right numbers for life."[55] It doesn't become statistically more likely. The unlikelihood of our fine-tuned universe remains unchanged. According to the calculation of a theoretical physicist Lee Smolin, the probability for our fine-tuned universe is one part in 10^{229}.[56] This is an insane probability, 1 in 10^{29} googol square.

In summary, multiverse is neither a logically viable nor probabilistically plausible solution of the fine-tuning problem. With zero evidence, and zero probability of observing another universe out of the multiverse, it shows how cosmologists, by their paradigm of inflation, have wondered into a cul-de-sac of non-science. It is not surprising because when the obvious solution of "7=7=7" is denied, intellectual dissonance will occur. To cosmologists, the origin of the universe remains a mystery.

As Steinhardt tells us, "the predictions of the early 1980s were based on a naive understanding of how inflation actually works—a picture that

54. Goff, "Improbable," para. 4.
55. Goff, "Improbable," para. 2.
56. Smolin, *Life*, 325.

has turned out to be dead wrong."[57] Since the inflation paradigm fails, the foundational problems in the plain vanilla Big Bang model which Alan Guth tried to solve remain unsolved.

How can we be assured of our safety when the Big Bang boat is ridden with holes?

57. Steinhardt, "Inflation," 41.

8. Conclusion

> He has made everything appropriate in its time. He has also set eternity in their heart, yet so that man will not find out the work which God has done from the beginning even to the end.[1]

8.1 Big Evolution

The picture painted by big evolutionists is popular among scientists and many people. According to Big Bang proponents, the origin of the universe happened about fourteen billion years ago. Big bio evolutionists say that the origin of life occurred about four billion years ago.

It seems that big evolution is omnipotent. Some cosmologists presume that the universe *could* begin from nothing. They believe that from no Nature, the laws of Nature could generate Nature. But is this belief logical?

We have at least two worldviews concerning the origins of various things around us. These two worldviews are incompatible. Although the observations are the same, the clash of worldviews is about two different historical accounts of origins. It is never about Christianity versus science.

Objection of big cosmic evolution is not against science. Defending big bio evolution is not a defense for science. Since origins are historical events, they can't be replicated, and they can't be observed. Therefore, with regard to origins, it is one version of history versus another one.

No two fields of science have the same goal. We need to distinguish fields that study how Nature operates today from those that speculate how Nature came yesterday. Therefore, it's germane to dub the former "operational science," and the latter "historical science."

Operational science is about repeatably testable cause and effect. It doesn't contradict the Bible. Operational science is consistent with the biblical worldview. Christians like Kepler wanted to bring glory and honor to God

1. Ecclesiastes 3:11.

by doing operational science. They were grateful for the privilege to study the wonders in Nature, knowing that they came by God's creation.

Once we acknowledge the existence of the supernatural realm, it is easy to understand that miracles can happen in God's timing. God, the Creator, knows how to suspend Nature at a precise place and time without causing lateral damages to the rest of the world. By virtue of God's unfathomable wisdom, every miracle He performs has a divine purpose. Since God is sovereign, no one can fully understand why He does or doesn't perform a miracle.

For the most part, Nature works as it was designed to operate. But after Adam fell, Nature became imperfect. Various diseases began to overwhelm all living things. Human life span also became shorter and shorter, from nine hundred years to ninety years. This is the biblical worldview. Genesis is a book about origins. Genesis means origin.

By contrast, the worldview of big evolution is that diseases are natural and that no human could possibly live for nine hundred years. Unlike operational science, historical science is about effects *to* their hypothetical cause. Knowing the effects, big evolutionists propose theories and initial conditions as candidates of origin. They use them to fit the effects they already know. Subsequent new observations either affirm their theories or contradict them.

Scientists working in the fields of historical science believe that big bio evolution is the historical cause of today's biodiversity in stasis. The jargon "stasis" refers to a state of equilibrium, i.e., no macroevolution. In stasis, monkeys today aren't evolving into some bipedal living things superior to humans. But in big evolutionism, there is no guarantee that they won't in the future if the equilibrium is "punctured."

On the other hand, nobody can deny microevolution. It is an observational science because we see these adaptive mutations happening every day. Nevertheless, most evolutionists recognize that many genetic mutations are neutral, some are harmful, and few are useful for adaptation. Genes are not the only determinant. Epigenetic modifications can shut down gene expression or switch it on. E. Coli still remain as bacteria, and fruit flies are still fruit flies. Also, artificial selection in controlled laboratories is different from natural selection in Nature. Most crucially, pre-existing gene pool is necessary for selection to take place.

These empirical facts make the unknown common ancestor questionable. It is said to have undergone unknown number of adaptive mutations over an unknown period of time. Extrapolated processes *were*, by construction, unobservable. Theories of evolution by extrapolation can't predict the life forms observed today and in the future.

In science, both the cause and the effect must be shown to be reproducible. Generally, a cause has several variables. We can observe the effect of one variable by varying its value while leaving the other variables unchanged. The purpose of such scientific experiments is to study how a controlled cause produces an observable effect.

Genetic mutations are controllable in the lab but not big evolution over dark deep time. Because historical events provide no variable for scientists to control, experiments are impossible for big evolution.

Similarly, Big Bang, big bounce, and the likes are theories in the sense of speculations. Big evolutionists postulate various historical causes of the unique Nature in which we live. Of course, the hypothesized causes themselves aren't repeatably testable by experiments.

If Nature today were to become unknown, theoreticians would have no way to fit their models with different initial conditions to the unknown final conditions we live in now.

Ancient philosophers had already speculated a lot on cosmology. They didn't have the kind of observations we have to fit their propositions to data. Surprisingly, some of the ancient proposals are quite similar to those of cosmologists today. In other words, we may consider historical science as a field of philosophical speculation about what had already happened.

Perhaps an easy way to understand why macroevolution and Big Bang are some philosophical assertions about the past is to politely ask a few simple questions like the following.

- Do you have a photo that captures the moment of Big Bang?
- Do you have a video of the formation process of the Milky Way?
- Do you have any video recording of star formation?
- Do you have any historical footage of the formation of Earth and Moon?
- Do you have the fossils of common ancestors?

These are fair questions. Frankly, the answers to these "yes" or "no" questions is "no." A common excuse for the absence of photos and videos is that the *processes* of big evolution took a very long time to complete. Essentially, the postulated causes are philosophical musings on what really happened.

On the other hand, operational science is oppositely different. We can video-record eclipses, comets, volcano eruptions, flower blossoms, ice formation, anti-particles, and more. We can repeat experiments on superconductivity, quantum tunneling, fractional quantum Hall effect, redshift, and the list goes on and on.

• • • • • •

We are told that Big Bang and macroevolution are science. Many people accept them as truths. Some are even prepared to defend them against critics. But these two theories are about the past, so big evolution is actually a God-less version of history.

To illustrate this point, suppose a sizable puddle is observed in an open field. A historical scientist builds a model of the past to "predict" the puddle. That is, a heavy rain occurred ten thousand years ago. Another historical scientist proposes a different model, saying that over a long period of time, meteorites brought water molecules to Earth little by little. Eventually, these water molecules formed the puddle. One particular model all big evolutionists avoid like a plague is this: Johnny the boy *intentionally* poured a lot of water last week to create the puddle.

This is how historical science works. Posit a *natural* cause such that it fits the effects we observe today. The argument is that because the model is consistent with observations, the historical cause and events this model can explain must be correct. On the face of it, this consistency argument is compelling. It is this connection to scientific observation that leads us to mistake Big Bang and macroevolution as facts.

The reality, however, is that the consistency of Big Bang with observations is one side of the coin. The other side is the inconsistency with the same observations. Understandably, this dark side is rarely mentioned in textbooks.

A major inconsistency lies in the "axis of evil" of the universe in the observational data known as the microwave background. The Big Bang theory assumes that space is isotropic, meaning that direction and location are completely irrelevant. The cosmos is supposed to look the same no matter which direction you turn to and from where you look. The microwave background is said to have a *cosmic* origin from the Big Bang. But microwave from the sky can be interpreted with other physical mechanisms that don't require a Big Bang.[2]

An axis in the microwave background is a special direction. It is an anomaly—a euphemism for the unwelcome surprise that the observation directly contradicts the assumption of isotropy.

What progress has been made on this thorny issue? Let us read a state-of-the-art review recently conducted by a team of 202 experts from fifty six academic research institutions around the world.[3] They have no vested in-

2. Ćirković and Perović, "Alternative," 1-18.

3. Abdalla et al., "Cosmology," 49–211.

terest in touting their own pet ideas to solve this conundrum. What is their concluding verdict? "No reasonable proposal has been made that would generate" the unexpected "axis of evil."[4] It's still there.

Even more surprising is the fact that the axis is aligned exactly with the elliptic plane of Earth orbiting the Sun, and the equinoxes of spring and autumn. In modern cosmology, such an alignment is crazy. When the cosmic microwave background came to be, the solar system didn't exist yet.

Moreover, we are told that the solar system is an insignificant speck in our galaxy, which in turn is an insignificant pebble in the universe. But the "axis of evil" takes this indoctrination to the janitor. Earth is, after all, at the center of the universe.

This Earth-centered interpretation is anathema to big evolutionists. The preferred scenario among cosmologists is that there is something wrong with the data and data analysis. But this is the same set of data that cosmologists use to support the Big Bang model. Different data analyses have been performed to cross-check, yet the "axis of evil" hasn't disappeared. What's going on?

Another major issue is the cosmological constant problem. Weinberg, an atheist and a Nobel laureate, concluded that "the cosmological constant is many orders of magnitude smaller than estimated in modern theories of elementary particles."[5] "The vacuum energy in the universe must be very small—about 120 orders of magnitude smaller than what quantum field theory predicts."[6] "It's generally regarded as one of the most awkward, embarrassing, difficult problems in theoretical physics today."[7]

If the actual measurement differs from the theoretical value by a thousand times, most scientists and engineers in other fields would think, "This theory really sucks. Let's throw it out of the window." A thousand times is 10^3; it is to write one followed by three zeros. This is miniscule compared to the discrepancy between the theoretical and actually measured values in the Big Bang cosmology. The order of magnitude of this discrepancy is one followed by 120 zeros:

> 100000000000000000000000000000000000000000000000000
> 000000000000000000000000000000000000000000000000000
> 0000000000000000000

4. Abdalla et al., "Cosmology," 130.
5. Weinberg, "Cosmological," 1.
6. Moskowitz, "Cosmic," para. 4.
7. Moskowitz, "Cosmic," para. 5.

In all my years as a researcher in different fields, I have never encountered such an astronomically *big* discrepancy. It is quite eye-opening.

That's not all. There are several "tensions" in mainstream Big Bang cosmology. The word "tension" is an encryption code to describe a shouldn't-happen phenomenon. Suppose you measure the temperature of water with different methods. All methods should give you the same value, up to a specified margin of error. This is common sense, as the temperature is not caused by the measurement method. If it were, science would become impossible.

Among several tensions in modern cosmology, a particularly deadly one is the measurement of the Hubble constant. It is *interpreted* as today's speed of expansion of the universe at large distances from Earth. This constant, denoted as H_0, is fundamental in determining the "age" of the universe. The reciprocal of H_0, or 1/ H_0, is *interpreted* in the Big Bang paradigm as the age of the universe.

Using the microwave background data *and* the standard Big Bang model, the Planck team reported that H_0 is 67.4 with a small error of plus minus 0.5. On the other hand, the supernova team led by Adam Reiss, a Nobel laureate, reported a statistically very different result. With an uncertainty of 1.3, their measurement of H_0 was 73.2. Thus, depending on which camp you support, the age of the universe is either 14.5 or 13.4 billion years, a difference of about one billion years.

A recent literature survey by top cosmologists describes the problem of H_0 tension as a "serious and desperate crisis in cosmology." "The significant discrepancy in the Hubble constant measurements . . . has raised a giant question mark" over the standard Big Bang model. "Despite its great success, the model is based on the assumption of three basic ingredients (cold dark matter, a cosmological constant, and inflation) whose underlying physics are largely unknown."[8]

Just recently, the most powerful James Webb Space Telescope in human history sent back the first batch of images. The digital images shocked the entire community of astronomers.

This telescope in space unveils galaxies furthest away from us. In the Big Bang paradigm, these galaxies should be small and primitive. But they look very similar to the familiar smooth discs and spiral galaxies near us. To make the tension worst, "there are up to ten times as many distant disk-shaped galaxies as was previously thought."[9] "The telescope's astonishingly

8. All the quotes in this paragraph are from Di Valentino et al., "Realm," 83.

9. Witze, "Four," 19.

sharp pictures have shattered astronomers' preconceptions about the early Universe."[10]

This new discovery "contradicts earlier theories of galaxy evolution," says Kirkpatrick, an astronomer. "Right now I find myself lying awake at three in the morning, wondering if everything I've ever done is wrong."[11]

Therefore, we have scientific grounds for doubting the Big Bang evolution. As more and better observations accumulate, cosmologists are now more forthcoming in revealing the dark secrets of Big Bang. Some of the cosmologists began to question, "Did Big Bang really happen?"

8.2 Biblical Creation

Christians are often criticized for not providing any *scientific* evidence for their claim that the Bible is divinely inspired. The skeptics of Christianity assert that the Bible is merely a book and that there is no such thing as divine inspiration.

Applying a data science method, I set up a statistical test with the null hypothesis that the signal of divine inspiration is zero, i.e., no signal. This statistical hypothesis reflects skeptics' view of the Bible. They claim that their zero-signal view is correct.

How can we test this null hypothesis versus the alternative hypothesis that the signal of divine inspiration is non-zero?

No mortal was around to witness or experience the creation and the Creator *in the beginning*. So whatever the Bible says about God and creation must be either a make-up story or divinely inspired. Thus, we focus on two God-creation verses that begin with "In the beginning." Let's test them.

Using two numerical schemes invented by the ancients, we map Genesis 1:1 in Hebrew and John 1:1 in Greek into numbers as our data. From these data, amazingly, two very important constants of math and one constant of quantum physics can be obtained with two simple but non-trivial formulas. At five significant digits, they are pi, the natural base of logarithms (Euler's number), and the fine structure constant (the mysterious 137).[12]

These three extraordinary numbers are very important and ubiquitous in science. Einstein's field equations of general relativity and quantum physics require them. If the biblical creation were a fiction, we would only have

10. Witze, "Four," 18.

11. Witze, "Four," 19.

12. I didn't discover this triplet signal. Most likely, Vernon Jenkins and his collaborators are the original discoverers. I merely replicated their calculations. I obtained exactly the same results. This data science is reproducible and repeatably testable.

random numbers. How can these constants of science pop out from two opening verses about the creation of the *physical* world and the Creator?

There is no *natural* reason for these three constants to be derived from the Bible. The authors Moses and John obviously didn't know these three numbers. But today we know that these numbers are heavily used in math and science.

They are like the signature on His masterpiece. Indeed, as Galileo quipped, "The universe . . . is written in the language of mathematics."[13]

• • • • • •

There are many ways to compose a sentence in Hebrew and Greek that don't change the meanings. For example, Moses could have written "the LORD God" rather than "God" in Hebrew for Genesis 1.1. If he did so, we wouldn't obtain pi. Similarly, if John added a definite article for "beginning" in Greek, then we wouldn't find the Euler number from John 1:1 either. And it follows that we wouldn't obtain the fine structure constant as well.

The probability of this set of three numbers being a fluke is one in more than seven trillion. It is the chance of picking a specially marked dime among all the dimes that cover the entire North Dakota. This improbability leaves us with only one option. We have a strong signal for divine inspiration.

Scientists don't doubt the discovery of the Higgs particle based on one in a few million. Therefore, scientists shouldn't doubt the discovery of the triplet signal for clearing the stringent threshold by a long shot.

Our data analysis firmly rejects the null hypothesis that the Bible has no signal of divine element. The trio of constants, which are frequently needed in science, is consistent with the literal theme of our data. God created the physical universe. He gave the laws of Nature. He knows all the math and science. So He knows how to suspend the laws. No animal is interested in the laws. We are made in His image and so we have the natural desire to find out the laws He gave.

The two verses reveal the Creator and His creation *in the beginning*. Nothing could be further from the truth that the Bible is just another book written by men. The scientific evidence is compelling.

The "Word" of John 1:1 in the context of John refers to the pre-incarnate Jesus, the trinitarian Creator. Our statistical analysis provides scientific support that Jesus is God the Creator. Since God doesn't lie, what Jesus said must be absolutely reliable and true.

13. Drake, *Discoveries*, 237–238.

Jesus affirmed the entire Old Testament. He also promised His disciples that the Holy Spirit would remind them of the words He taught. Mark was a close associate of Peter, and Luke of Paul. Thus, the twenty seven books of the New Testament are also divinely inspired.

Christianity is rational and verifiable. Either the tomb of Jesus was empty or it wasn't. To argue for this case, we must first know that lying is a sin, as it breaks one of the ten commandments: Thou shall not bear false witness. Now, Paul, a pious Pharisee, a strict Law abider, had a close encounter with Jesus. He was dramatically transformed from a persecutor of God's church to an apostle. He wrote that Jesus "appeared to more than five hundred brethren at one time, most of whom remain until now, but some have fallen asleep."[14]

If Jesus had not been resurrected and alive, there wouldn't be any apostle and Christianity would be gone forever in the first century.

We have many other reasons to believe that Jesus's resurrection is a historical fact. The New Testament contains the words of Jesus and obviously His words have the divine origin. Furthermore, the Holy Spirit helped the New Testament writers to teach, correct, encourage, and comfort Christians who were persecuted by the Jewish leaders and the Roman government in the first century. They had no reason to write a fiction.

• • • • • •

Jesus affirmed the literal interpretation of Genesis. So did Moses and the other prophets. They were men of God who preached the words of the LORD. The disciples and followers of Jesus in the first century believed what Jesus said about creation.

Secular scientists describe the Big Bang cosmology as being "in a crisis." Nobody knows how cosmology will pen out in the future. But the words of Jesus are always true. Jesus is the One who created the universe and everything in it. Christians can't afford *not* to trust what Jesus said about the creation. In light of the fatal scientific flaws in big evolution, isn't it unwise to place historical science above the divinely inspired Bible?

It makes more sense to believe in biblical creation than in big evolution.

• • • • • •

Finally, as a personal testimony, I found that my faith increased many folds since I began to trust Jesus's literal interpretation of Genesis. Every word of

14. 1 Corinthians 15:6.

the Bible has become precious to me. I also became more forthcoming to serve in the church, and I consider serving as a privilege.

And in this way God loved the world: "As Moses lifted up the serpent in the wilderness, even so must the Son of Man be lifted up; so that whoever believes will in Him have eternal life."[15] And God "gave His only begotten Son, that whoever believes in Him shall not perish, but have eternal life."[16]

Jesus, the Creator, died on the lifted-up cross in our place so that our sins are forgiven. Jesus was "a man of sorrows and acquainted with grief."[17] "He was despised and forsaken of men."[18] Jesus was openly humiliated. He suffered bodily pain and emotional trauma. "He was pierced through for our transgressions. He was crushed for our iniquities. The chastening for our well-being fell upon Him, and by His scourging we are healed."[19]

Jesus "poured out Himself to death, and was numbered with the transgressors. Yet He Himself bore the sin of many, and interceded for the transgressors."[20]

This is the gospel prophesized by Isaiah many centuries before Jesus came as Immanuel. This is the gospel preached by Jesus and His disciples. All we need to do is to acknowledge sincerely that "All of us like sheep have gone astray. Each of us has turned to his own way."[21] We all wanted to be god with our own way. God will forgive when we repent, for "the LORD has caused the iniquity of us all to fall on"[22] Jesus.

• • • • • •

Jesus is God the Creator. Afterlife is real. God our Savior "desires all men to be saved and to come to the knowledge of the truth."[23] It just takes a little bit of humility to start reading the Bible, to pray and repent—a change of mind, heart, and actions.

15. John 3:14–15.
16. John 3:16.
17. Isaiah 53:3.
18. Isaiah 53:3.
19. Isaiah 53:5.
20. Isaiah 53:12.
21. Isaiah 53:6.
22. Isaiah 53:6.
23. 1 Timothy 2:4.

God is the perfect and final judge. Yet at the same time, God is also the forgiver in Jesus Christ. My prayer for you is that "you will know the truth, and the truth will make you free."[24]

24. John 8:32.

Bibliography

Abdalla, Elcio, et al. "Cosmology Intertwined: A Review of the Particle Physics, Astrophysics, and Cosmology Associated with the Cosmological Tensions and Anomalies." *Journal of High Energy Astrophysics* 34 (2022) 49–211. https://linkinghub.elsevier.com/retrieve/pii/S2214404822000179.

Ade, P. A. R., et al. "Planck 2015 results XIII. Cosmological parameters." *Astronomy & Astrophysics* 594 (2016) A13. https://doi.org/10.1051/0004-6361/201525830.

Adler, Ronald J., et al. "Vacuum Catastrophe: An Elementary Exposition of the Cosmological constant Problem." *American Journal of Physics* 63 (1995) 620–626. https://doi.org/10.1119/1.17850.

Aron, Jacob. "Planck Shows Almost Perfect Cosmos—Plus Axis of Evil." *NewScientist* (March 21, 2013).

Atiyah, M. F. "The Work of Edward Witten." In *Fields Medallists' Lectures*, edited by M. F. Atiyah and D. Iagolnitzer, 514–518. World Scientific Series in 20th Century Mathematics 5. River Edge: World Scientific, 1997.

ATLAS Collaboration. "Latest Results from ATLAS Higgs Search." *CERN Accelerating Science* (July 4, 2012). https://atlas.cern/updates/press-statement/latest-results-atlas-higgs-search.

Bak, Per. *How Nature Works: The Science of Self-Organized Criticality*. New York: Springer-Verlag, 1996.

Batten, Don. "'It's not Science.'" *Creation Ministries International* (September 18, 2014). https://creation.com/its-not-science.

Bergman, Jerry. *Hitler and the Nazi Darwinian Worldview*. Kitchener: Joshua Press, 2012.

Berlinguette, Curtis P., et al. "Revisiting the Cold Case of Cold Fusion." *Nature* 570 (2019) 45–51.

Blei, David M., and Padhraic Smyth. "Science and Data Science." *Proceedings of the National Academy of Sciences of the United States of America* 114 (2017) 8689–8692.

Block, Martin M., and Francis Halzen. "Experimental Confirmation that the Proton is Asymptotically a Black Disk." *Physics Review Letters* 107 (2011) 212002 1–4. https://link.aps.org/doi/10.1103/PhysRevLett.107.212002.

Britannica. "Sir Fred Hoyle." *Encyclopedia Britannica*, (March 15, 2022). https://www.britannica.com/biography/Fred-Hoyle. Accessed 30 July 2022.

———. "Alfonsine Tables." *Encyclopedia Britannica* (June 21, 2002). https://www.britannica.com/topic/Alfonsine-Tables.

Callaway, Ewen. "Legendary Bacterial Evolution Experiment Enters New Era." *Nature* 606 (2022) 634–635.

CERN. "Higgs within Reach." *CERN News* (July 4, 2012). https://home.cern/news/news/physics/higgs-within-reach.

———. "CERN Experiments Announce First Indications of a Rare Higgs Boson Process." *CERN News* (August 3, 2020). https://home.cern/news/press-release/physics/cern-experiments-announce-first-indications-rare-higgs-boson-process.

CFI Team. "Qualitative Characteristics of Accounting Information: The Fundamental (Primary) and Enhancing (Secondary) Qualitative Characteristics." *CFI* (April 22, 2022).

Chow, Tai L. *Gravity, Black Holes, and the Very Early Universe: An Introduction to General Relativity and Cosmology*. New York: Springer, 2008.

Ćirković, Milan M., and Slobodan Perović. "Alternative Explanations of the Cosmic Microwave Background: A Historical and an Epistemological perspective." *Studies in History and Philosophy of Modern Physics* 62 (2018) 1–18. https://doi.org/10.1016/j.shpsb.2017.04.005.

Cleaves, H. James, et al. "A Reassessment of Prebiotic Organic Synthesis in Neutral Planetary Atmospheres." *Origins of Life and Evolution of Biospheres* 38 (2008) 105–115.

CMS Experiment. "Observation of a New Particle with a Mass of 125 GeV." *CERN* (July 4, 2012). https://cms-docdb.cern.ch/cgi-bin/PublicDocDB/RetrieveFile?docid=6116&filename=CMShiggs2012_EN.pdf.

CNN reporter. "Finland in Mourning after Fatal School Shooting." *CNN* (November 8,, 2007). https://edition.cnn.com/2007/WORLD/europe/11/08/school.shooting/index.html.

Confucius Institute for Scotland. "1, 2, 3—the Legend of Pangu." *Confucius Institute for Scotland in the University of Edinburgh*. https://www.confuciusinstitute.ac.uk/1-2-3-the-legend-of-pangu/.

Conover, Emily, "Edward Witten Receives Inaugural APS Medal." *APS News* (February 3, 2016). https://www.aps.org/publications/apsnews/updates/witten.cfm.

Cooperman, Alan, et al. "U.S. Public Becoming Less Religious: Modest Drop in Overall Rates of Belief and Practice, but Religiously Affiliated Americans Are as Observant as Before." *Pew Research Center* (November 3, 2015). https://www.pewresearch.org/religion/wp-content/uploads/sites/7/2015/11/201.11.03_RLS_II_full_report.pdf.

Copernicus, Nicolaus. *On the Revolutions of the Heavenly Spheres*. Translated by Charles Gen Wallis. Nuremberg: Johannes Petreius, 1543.

Corneliussen, Steven T. "Cosmic Inflation Debate Bleeds into Popular Science Media." *Physics Today* (June 5, 2017). https://physicstoday.scitation.org/do/10.1063/pt.6.3.20170605a/full/.

Couronne, Ivan, and Issam Ahmed. "Top Cosmologist's Lonely Battle against 'Big Bang' Theory." *Phys.Org* (November 14, 2019). https://phys.org/news/2019-11-cosmologist-lonely-big-theory.html.

Coyne, Jerry A. *Why Evolution is True*. Oxford: Oxford University Press, 2009.

———. "An Attack on Evolution, Disguised as a Darwin Biography." *The Washington Post* (December 28, 2017). https://www.washingtonpost.com/outlook/an-attack-on-evolution-disguised-as-a-darwin-biography/2017/12/28/ce3df238-b502-11e7-9e58-e6288544af98_story.html.

Crick, Francis. *Life Itself: Its Origin and Nature*. New York: Simon and Schuster, 1981.

———. *What Mad Pursuit: A Personal View of Scientific Discovery*. New York: HarperCollins, 1988.

———. *The Astonishing Hypothesis: The Scientific Search for the Soul*. New York: Charles Scribner's Sons, 1994.

Cyburt, Richard H., et al. "An Update on the Big Bang Nucleosynthesis Prediction for ^{7}Li: the Problem Worsen." *Journal of Cosmology and Astroparticle Physics* 2008 (2008) 012. https://doi.org/10.1088/1475-7516/2008/11/012.

Davis, Ted. "Theistic Evolution: History and Beliefs." *Biologos* (October 15, 2018). https://biologos.org/series/science-and-the-bible/articles/theistic-evolution-history-and-beliefs.

da Vinci, Leonardo. *Notebooks*. Oxford World's Classics. Selected by Irma A. Richter; Edited with an introduction and notes by Thereza Wells; Preface by Martin Kemp. Oxford: Oxford University Press, 2008.

Dawkins, Richard. "Is it a Theory? Is it a Law? No, it's a fact." *Richard Dawkins Foundation for Reason and Science* (November 30, 2015). https://richarddawkins.net/2015/11/is-it-a-theory-is-it-a-law-no-its-a-fact/.

Deibert, Emily. "How much stuff is there in the universe?" *NewScientist* (August 20, 2020). https://www.newscientist.com/lastword/mg24732961-500-how-much-stuff-is-there-in-the-universe/.

Deprit, A. "Monsignor Georges Lemaître." In *The Big Bang and Georges Lemaitre: Proceedings of a Symposium in Honour of G. Lemaître Fifty Years after His Initiation of Big-Bang Cosmology, Louvain-Ia-Neuve, Belgium, 10–13 October 1983*, edited by A. Berger, 363–392. Dordrecht: D. Reidel Publishing, 1984.

de Santillana, Giorgio. *The Crime of Galileo*. Chicago: University of Chicago Press, 1976.

DiChristina, Mariette. "A Cosmic Controversy." *Scientific America* (May 10, 2017). https://blogs.scientificamerican.com/observations/a-cosmic-controversy/.

Di Valentino, Eleonora, et al. "Planck Evidence for a Closed Universe and a Possible Crisis for Cosmology." *Nature Astronomy* 4 (2020) 196–203. https://www.nature.com/articles/s41550-019-0906-9.

———. "In the Realm of the Hubble Tension—a Review of Solutions." *Classical and Quantum Gravity* 38 (2021) 153001. https://iopscience.iop.org/article/10.1088/1361-6382/ac086d.

Dirac, Paul "The Quantum Theory of the Electron." *Proceedings of the Royal Society A* 117 (1928) 619–624. https://doi.org/10.1098/rspa.1928.0023.

———. "The Evolution of the Physicist's Picture of Nature." *Scientific American* 208 (1963) 45–53. http://www.jstor.org/stable/24936146.

———. "Methods in Theoretical Physics." In *From a Life of Physics: Evening Lectures at the International Centre for Theoretical Physics, Trieste, Italy*. International Atomic Energy Agency, 1969.

Dobzhansky, Theodosius. "Nothing in Biology Makes Sense except in the Light of Evolution." *The American Biology Teacher* 35 (1973) 125–129. https://doi.org/10.2307/4444260.

Dose, Klaus. "The Origin of Life: More Questions than Answers." *Interdisciplinary Science Reviews* 13 (1988). 348–356.

Drake, Stillman. *Discoveries and Opinions of Galileo*. New York: Doubleday Anchor Book, 1957.

Englert, F., and R. Brout. "Broken Symmetry and the Mass of Gauge Vector Mesons." *Physical Review Letters* 13 (1964) 321–323. https://link.aps.org/doi/10.1103/PhysRevLett.13.321.

Einstein, Albert. "Comment on A. Friedmann's Paper: 'On the Curvature of Space.'" In *The Berlin Years: Writings & Correspondence, January 1922 - March 1923*, edited by Diana Kormos Buchwald et al. Translated by Ann M. Hentschel and Osik Moses, 271–272. The Collected Papers of Albert Einstein 13. Princeton: Princeton University Press, 2012. https://einsteinpapers.press.princeton.edu/vol13-trans/302.

Farbridge, Maurice Henry. *Studies in Biblical and Semitic Symbolism*. London: K. Paul, Trench, Trubner & Co, 1923.

Ferguson, Andrew. "The Heretic." *The Weekly Standard* (March 25, 2013). https://www.washingtonexaminer.com/weekly-standard/the-heretic.

Ferguson, Robert. *The Penalties of Greatness*. London: Ward and Co, 1858. https://play.google.com/books/reader?id=2SYFAAAAIAAJ&pg=GBS.PP8&hl=en.

Feynman, Richard. *QED: The Strange Theory of Light and Matter*. Princeton: Princeton University Press, 1988.

———. *The Meaning of it All: Thoughts of a Citizen-Scientist*. Reading: Perseus Books, 1998.

Feynman, Richard, et al. *The Feynman Lectures on Physics, Volume 1*. Online edition. California Institute of Technology, 2013. https://www.feynmanlectures.caltech.edu/I_toc.html.

Fields, Brian D., et al. "Big-Bang Nucleosynthesis after Planck." *Journal of Cosmology and Astroparticle Physics* 2020 (2020) 010.

Finocchiaro, Maurice A. *The Galileo Affair: A Documentary History California Studies in the History of Science*. Berkeley: University of California Press, 1989.

Fleischmann, Martin, and Stanley Pons. "Electrochemically Induced Nuclear Fusion of Deuterium." *Journal of Electroanalytical Chemistry and Interfacial Electrochemistry* 261 (1989) 301–308.

Forbes, Andrew A., and Billy A. Krimmel. "Evolution Is Change in the Inherited Traits of a Population through Successive Generations." *Nature Education Knowledge* 3(10) (2010) 6. https://www.nature.com/scitable/knowledge/library/evolution-is-change-in-the-inherited-traits-15164254/.

Friedlander, Michael Wulf. "Cosmic Ray." *Encyclopaedia Britannica* (September 27, 2021). https://www.britannica.com/science/cosmic-ray/Studying-cosmic-rays.

Friedmann, Aleksandr. "On the Curvature of Space." Translated by Brian Doyle. *Zeitschrift für Physik* 10 (1922) 377–386. https://wwwphy.princeton.edu/~steinh/ph563/friedmann.pdf.

Galileo. "Letter to Benedetto Castelli." *Works by Galileo Galilei* (December 21, 1613). https://inters.org/Galilei-Benedetto-Castelli.

Gamow, George. *My World Line: An Informal Autobiography*. New York: Viking, 1970.

Gillen, Alan, and Frank Sherwin. "Louis Pasteur's Views on Creation, Evolution, and the Genesis of Germs." *Answers Research Journal* 1 (2008) 43–52. https://answersresearchjournal.org/pasteur-creation-evolution-germs/.

Gingerich, Owen. *The Book Nobody Read: Chasing the Revolutions of Nicolaus Copernicus*. New York: Walker & Company, 2004.

Goff, Philip. "Our Improbable Existence Is No Evidence for a Multiverse." *Scientific American* (January 10, 2021). https://www.scientificamerican.com/article/our-improbable-existence-is-no-evidence-for-a-multiverse/.

Gould, Stephen Jay. *Ever Since Darwin: Reflections in Natural History*. New York: W. W. Norton & Company, 1977.

Green, J. A. M., et al. "Explicitly Modelled Deep-Time Tidal Dissipation and Its Implication for Lunar History." *Earth and Planetary Science Letters* 461 (2017) 46–53.

Gregory, Andrew. *Ancient Greek Cosmogony*. London: Bloomsbury, 2007.

———. "Description." *Ancient Greek Cosmogony* (2008). https://www.bloomsbury.com/us/ancient-greek-cosmogony-9780715634776/.

Guliuzza, Randy J. *Twenty Evolutionary Blunders: Dangers and Difficulties of Darwinian Thinking*. Texas: Institute for Creation Research, 2017.

Guralnik, G., et al. "Global Conservation Laws and Massless Particles." *Physical Review Letters* 13 (1964) 585–587. https://link.aps.org/doi/10.1103/PhysRevLett.13.585.

Guth, Alan H. "Inflation and the New Era of High-Precision Cosmology." *MIT Physics Annual* (2002) 28–39. https://physics.mit.edu/wp-content/uploads/2021/01/physicsatmit_02_cosmology.pdf.

———. "Eternal Inflation and its Implications." *Journal of Physics A: Mathematical and Theoretical* 40 (2007) 6811–6826. https://doi.org/10.1088/1751-8113/40/25/S25.

Gunn, Alastair. "How Dense is the Universe?" *BBC Science Focus Magazine*. https://www.sciencefocus.com/space/how-dense-is-the-universe/.

Haeckel, Ernst. *The History of Creation, Vol. 1*. Translated by E. Ray Lankester. Salt Lake City: Project Gutenberg, 2012. https://www.gutenberg.org/files/40472/40472-h/40472-h.htm.

Hawking, Stephen. *A Brief History Of Time: From the Big Bang to Black Holes*. Updated version. London: Transworld, 2009.

———. *Brief Answers to the Big Questions*. New York: Random House, 2018.

Heilbron, John L. *Galileo*. Oxford: Oxford University Press, 2010.

Higgs, Peter W. "Broken Symmetries and the Masses of Gauge Bosons." *Physical Review Letters* 13 (1964) 508–509. https://link.aps.org/doi/10.1103/PhysRevLett.13.508.

Hobbs, William H. "The Making of Scientific Theories." *Science* 45 (1917) 441–451.

Hobson, M. P., et al. *General Relativity: An Introduction for Physicists*. Cambridge: Cambridge University Press, 2006.

Horgan, John. "Physicist Slams Cosmic Theory He Helped Conceive." *Scientific American* (December 1, 2014). https://blogs.scientificamerican.com/cross-check/physicist-slams-cosmic-theory-he-helped-conceive/.

———. *The End of Science: Facing the Limits of Knowledge in the Twilight of the Scientific Age*. New York: Basic Books, 2015.

Hossenfelder, Sabine. "Why the Multiverse is Religion, not Science." *Backreaction* (July 9, 2019). http://backreaction.blogspot.com/2019/07/why-multiverse-is-religion-not-science.html.

Hubble, Edwin. "A Relation between Distance and Radial Velocity among Extra-Galactic Nebulae." *Proceedings of the National Academy of Sciences* 15 (1929) 168–173. https://doi.org/10.1073/pnas.15.3.168.

———. *The Observational Approach to Cosmology*. Oxford: Clarendon, 1937. http://ned.ipac.caltech.edu/level5/Sept04/Hubble/Hubble_contents.html.

Huizenga, John Robert. *Cold Fusion: The Scientific Fiasco of the Century*. New York: University of Rochester Press, 1992.

IBM Cloud Education. "What is Data Science?" *IBM Cloud Learn Hub* (May 15, 2020). https://www.ibm.com/cloud/learn/data-science-introduction.

ICTP Staff. "Dirac Medallists 2002." *The Medallists* (2002). https://www.ictp.it/about-ictp/prizes-awards/the-dirac-medal/the-medallists/dirac-medallists-2002.aspx.

Ifrah, Georges. *The Universal History of Numbers: From Prehistory to the Invention of the Computer*. Translated by Bellos et al. New York: John Wiley & Sons, 2000.

Ijjas, Anna, et al. "Inflationary Paradigm in Trouble after Planck2013." *Physics Letters B* 723 (2013) 261–266. https://doi.org/10.1016/j.physletb.2013.05.023.

———. "Pop Goes the Universe." (2017). *Scientific American* 316 (February 2017) 32–39.

———. "Pop Goes the Universe." (2017). https://physics.princeton.edu//~cosmo/sciam/index.html.

Iliffe, Robert. "Newton's Religious Life and Work." *The Newton Project* (2013). https://www.newtonproject.ox.ac.uk/view/contexts/CNTX00001.

Jang, Juhi, and Nader Masmoudi. "Derivation of Ohm's Law from the Kinetic Equations." *SIAM Journal on Mathematical Analysis* 44 (2012) 3649–3669. https://www.proquest.com/openview/55bd19d5a4e2d4d75aa3ad28dd875abc/1?pq-origsite=gscholar&cbl=666308.

Jeanson, Nathaniel T. *Replacing Darwin: The New Origin of Species*. Green Forest: Master Books, 2017.

Jenkins, Jere H., et al. "Evidence of Correlations between Nuclear Decay Rates and Earth–Sun Distance." *Astroparticle Physics* 32 (2009) 42–46. https://doi.org/10.1016/j.astropartphys.2009.05.004.

Jenkins, Vernon. "The Ultimate Assertion: Evidence of Supernatural Design in the Divine Prologue." *Journal of Creation* 7 (1993) 184–196. https://creation.com/images/pdfs/tj/j07_2/j07_2_184-196.pdf.

———. "Themes from Creation's Blueprint Part 2: Three Fundamental Constants." (2003). http://www.whatabeginning.com/Themes/Part2/BP.htm.

———. *The Second Edge: A Role for Numerical Coincidence in the Pursuit of Truth*. Dublin: Mentor, 2016.

Johnson, Chris, et al. "Earth History." In *An Introduction to Geology*. Free Textbook for College-Level Introductory Geology Courses. Salt Lake Community College, 2017. https://opengeology.org/textbook/8-earth-history/.

Jow, Dylan L., and Douglas Scott. "Re-Evaluating Evidence for Hawking Points in the CMB." *Journal of Cosmology and Astroparticle Physics* 2020 (2020) 021–021. https://doi.org/10.1088/1475-7516/2020/03/021.

Keas, Michael Newton. *Unbelievable: 7 Myths About the History and Future of Science and Religion*. Wilmington: ISI Books, 2019.

Kepler, Johannes. *Epitome of Copernican Astronomy & Harmonies of the World*. Translated by Charles Glenn Wallis. New York: Prometheus Books, 1995.

———. *The Harmony of the World*. Translated by E. J. Aiton et al. Philadelphia: American Philosophical Society, 1997.

Knapp, Alex. "How Much Does it Cost to Find a Higgs Boson?" *Forbes* (July 5, 2012). https://www.forbes.com/sites/alexknapp/2012/07/05/how-much-does-it-cost-to-find-a-higgs-boson/?sh=542f85b43948.

Krauss, Lawrence M. "The Energy of Empty Space That Isn't Zero." *Edge.org* (July 5, 2006). https://www.edge.org/conversation/lawrence_m_krauss-the-energy-of-empty-space-that-isnt-zero.

———. *A Universe from Nothing: Why There Is Something Rather than Nothing*. New York: Free, 2012.

———. "The Stars Died So That You Could Be Here Today." *teosTV*, YouTube channel, 2012.

Kunimoto, Michelle, et al. "Searching the Entirety of *Kepler* Data. I. 17 New Planet Candidates Including One Habitable Zone World." *Astronomical Journal* 159:124 (2020) 1–20. https://doi.org/10.3847/1538-3881/ab6cf8.

Land, Kate, and João Magueijo. "Examination of Evidence for a Preferred Axis in the Cosmic Radiation Anisotropy." *Physics Review Letters* 95 (2005) 071301. https://journals.aps.org/prl/abstract/10.1103/PhysRevLett.95.071301.

Lederman, Leon M., and Dick Teresi. *The God Particle: If the Universe Is the Answer, What Is the Question?* New York: Houghton Mifflin Harcourt Publishing, 2006.

Lemaître, Abbé G. "A Homogeneous Universe of Constant Mass and Increasing Radius accounting for the Radial Velocity of Extra-galactic Nebulae." *Monthly Notices of the Royal Astronomical Society* 91 (1931) 483–490. https://doi.org/10.1093/mnras/91.5.483.

Lennox, John C. *God's Undertaker: Has Science Buried God?* Oxford: Lion Hudson, 2007.

Lewis, C. S. *Miracles: A Preliminary Study*. New York: Simon & Schuster, 1996.

Lewontin, Richard. "Billions and Billions of Demons." *The New York Review* (January 7, 1997).

Lisle, Jason. *Why Genesis Matters: Christian Doctrine and the Creation Account*. Dallas: Institute for Creation Research, 2012,

Literary Digest Writer. "Is Darwinism on Its Death-Bed." *Literary Digest* 25 (October 18, 1902).

Loeb, Abraham. "A Scientist Must Go Where the Evidence Leads." *Scientific America Blog* (August 14, 2019). https://blogs.scientificamerican.com/observations/a-scientist-must-go-where-the-evidence-leads/.

Marmet, L. "On the Interpretation of Spectral Red-Shift in Astrophysics: A Survey of Red-Shift Mechanisms—II." *arXiv e-prints* (2018) arXiv:1801.07582.

Martin, Joel W. "Compatibility of Major U.S. Christian Denominations with Evolution." *Evolution: Education and Outreach* 3 (2010) 420–431.

Masci, David. "For Darwin Day, 6 Facts about the Evolution Debate." *Pew Research Center* (February 11, 2019). https://www.pewresearch.org/fact-tank/2019/02/11/darwin-day/.

Masters, Karen. "As the Universe Expands, Why Don't Galaxies Get Stretched Out? (Intermediate)." *Ask an Astronomer* (June 27, 2015). http://curious.astro.cornell.edu/about-us/97-the-universe/galaxies/cosmology/538-as-the-universe-expands-why-don-t-galaxies-get-stretched-out-intermediate.

Maxwell, James Clerk. "Molecules." In *The scientific papers of James Clerk Maxwell Volume* 2, edited by W. D. Niven, 376. Cambridge: Cambridge University Press, 2011.

Merritt, Cam. What Is Verifiability in Accounting?" *Chron* (April 10, 2019). https://smallbusiness.chron.com/verifiability-accounting-56715.html.

Meyer, Stephen C. *Darwin's Doubt: The Explosive Origin of Animal Life and the Case for Intelligent Design*. London: HarperCollins Publishers, 2013.

Migkas, K., et al. "Probing Cosmic Isotropy with a New X-ray Galaxy Cluster Sample through the L_X-T Scaling Relation." *Astronomy & Astrophysics* 636 (2020) A15. https://doi.org/10.1051/0004-6361/201936602.

Miller, Lisa, et al. "Neuroanatomical Correlates of Religiosity and Spirituality: A Study in Adults at High and Low Familial Risk for Depression." *JAMA Psychiatry* 71 (2014) 28–135. https://www.ncbi.nlm.nih.gov/pmc/articles/PMC3921896/.

Moskowitz, Clara. "Physicists Disagree Over Meaning of Quantum Mechanics, Poll Shows." *Live Science* (January 21, 2013). https://www.livescience.com/26444-quantum-mechanics-physicists-poll.html.

———. "Cosmic Conundrum." *Scientific American* 324 (February 2021) 24–29. https://www.scientificamerican.com/article/the-cosmological-constant-is-physics-most-embarrassing-problem/.

Musha, Takaaki. "Alternative Explanation of the Cosmological Red Shift by the Tachyon Plasma Field in Intergalactic Space." *Global Journal of Science Frontier Research: (A) Physics and Space Science* 20 (2020) 26–31. https://journalofscience.org/index.php/GJSFR/article/download/2812/2673.

Nadler, Steven. "Baruch Spinoza." *The Stanford Encyclopedia of Philosophy* (Summer 2022 Edition), edited by Edward N. Zalta. https://plato.stanford.edu/archives/sum2022/entries/spinoza.

Naeye, Robert. "Milky Way Supernova Rate Confirmed." *Sky & Telescope* (January 4, 2006). https://skyandtelescope.org/astronomy-news/milky-way-supernova-rate-confirmed/.

Nagel, Thomas. *Mind and Cosmos: Why the Materialist Neo-Darwinian Conception of Nature Is Almost Certainly False*. Oxford: Oxford University Press, 2012.

Naked Science. "Birth of the Earth." *Naked Science*, YouTube Channel, 2015. https://www.youtube.com/watch?v=XYHe5wQeA28.

National Park Service Staff. "Geologic Principles—Uniformitarianism." *National Park Service* (September 27, 2018). https://www.nps.gov/articles/geologic-principles-uniformitarianism.htm.

National Post Staff. "'What has gotten into Thomas Nagel?': Leading Atheist Branded a 'Heretic' for Daring to Question Darwinism." *National Post* (March 23, 2013). https://nationalpost.com/holy-post/what-has-gotten-into-thomas-nagel-leading-atheist-branded-a-heretic-for-daring-to-question-darwinism.

Newport, Frank. "Fewer in U.S. Now See Bible as Literal Word of God." *Gallup* (July 5, 2022). https://news.gallup.com/poll/394262/fewer-bible-literal-word-god.aspx.

Newton, Isaac. "A Short Schem of the True Religion." *The Newton Project* (February 2002). https://www.newtonproject.ox.ac.uk/view/texts/normalized/THEM00007.

Nietzsche, Friedrich. *The Gay Science*. Translated by Walter Kaufmann. New York: Random House, 1974.

O'Connor, J. J., and E. F. Robertson. "Aleksandr Aleksandrovich Friedmann." *MacTutor* (December 1997). https://mathshistory.st-andrews.ac.uk/Biographies/Friedmann/.

Pagel, Kymberleigh A., et al. "When Loss-of-function is Loss of Function: Assessing Mutational Signatures and Impact of Loss-of-Function Genetic Variants." *Bioinformatics* 33 (2017) i389–i398. https://doi.org/10.1093/bioinformatics/btx272.

Pagels, Heinz R. *The Cosmic Code: Quantum Physics as the Language of Nature*. New York: Bantam Book, 1982.

Pais, Abraham. "Paul Dirac: Aspects of his Life and Work." In *Paul Dirac: the Man and his Work*, edited by Peter Goddard, 1–45. Cambridge: Cambridge University Press, 1998.

Panek, Richard. "A Cosmic Crisis." *Scientific American* 322 (March 2020) 30–37. https://www.scientificamerican.com/article/how-a-dispute-over-a-single-number-became-a-cosmological-crisis/.

Penrose, Roger. *Cycles of Time: An Extraordinary New View of the Universe*. London: The Bodley Head, 2010.

Petit, Jean-Pierre. "An Interpretation of Cosmological Model with Variable Light Velocity." *Modern Physics Letters A* 3 (1988) 1527–1532.

Planck Collaboration. "Planck's New Cosmic Recipe." Updated September 2019. *European Space Agency News Archive* (March 20, 2013). https://sci.esa.int/web/planck/-/51557-planck-new-cosmic-recipe.

Powell, Mark Allan. *Introducing the New Testament: A Historical, Literary, and Theological Survey*. 2nd ed. Ada: Baker Academic, 2018. http://bakerpublishinggroup.com/books/introducing-the-new-testament-2nd-edition/11940/students/esources.

Pross, Andy, and Robert Pascal. "The Origin of Life: What we Know, What we can Know and What we will Never Know." *Open Biology* 3 (2013) 120190. https://royalsocietypublishing.org/doi/10.1098/rsob.120190.

Provine, William B., and Phillip E. Johnson. "Darwinism: Science or Naturalistic Philosophy? A Debate between William B. Provine and Phillip E. Johnson at Stanford University, April 30, 1994." *Access Research Network* 16 (1995). http://www.arn.org/docs/orpages/or161/161main.htm.

Rahman, Syed Faisal ur. "The Enduring Enigma of the Cosmic Cold Spot." *Physics World* 33 (2020) 36–40. https://doi.org/10.1088/2058-7058/33/2/35.

Reichert, Stefanie. "e is everywhere." *Nature Physics* 15 (2019) 982–982. https://doi.org/10.1038/s41567-019-0655-9.

Rennie, John. "15 Answers to Creationist Nonsense." *Scientific American* 287 (2002). https://www.scientificamerican.com/article/15-answers-to-creationist/.

Robertson, Edmund, and John O'Connor. "Quotations: Isaac Newton." *MacTutor*. https://mathshistory.st-andrews.ac.uk/Biographies/Newton/quotations/

Rogers, Kara. "Abiogenesis." *Encyclopedia Britannica* (March 15, 2022). https://www.britannica.com/science/abiogenesis. Accessed September 26, 2022.

Ruse, Michael. "Is Evolution a Secular Religion?" *Science* 299 (2003) 1523–1524. https://www.science.org/doi/10.1126/science.1082968.

Sagan, Carl. *The Cosmic Connection: An Extraterrestrial Perspective*. Produced by Jerome Agel. New York: Dell Publishing, 1973.

Saeki, Kimiko, et al. "New Method to Prepare Single-Headed Heavy Meromyosin with High Purity and a High Yield." *Journal of Biochemistry* 106 (1989) 606–611. https://www.jstage.jst.go.jp/article/biochemistry1922/106/4/106_4_606/_pdf/-char/en.

Sarfati, Jonathan. *The Greatest Hoax on Earth? Refuting Dawkins on Evolution*. Brisbane: Creation Book, 2010.

Sarfati, Jonathan, and Mike Matthews. *Refuting Evolution 2*. Green Forest: Master Book, 2002.

Sheffield, Sarah L., and Jennifer E. Bauer. "Darwin Day in Deep Time: Promoting Evolutionary Science through Paleontology." *Evolution Education Outreach* 10, 10 (2017). https://doi.org/10.1186/s12052-017-0073-3.

Schleicher, Andreas. "PISA 2018: Insights and Interpretations." *OECD*, 2019.

Sewell, Keith C. *Herbert Butterfield and the Interpretation of History.* New York: Palgrave Macmillan, 2005.

Smolin, Lee. *The Life of the Cosmos.* Oxford: Oxford University Press, 1997.

Sommer, Joseph C. "Some Reasons Why Humanists Reject The Bible." *American Humanist Association* (undated). https://americanhumanist.org/what-is-humanism/reasons-humanists-reject-bible/.

Spence, Joseph. *Observations, Anecdotes, and Characters, of Books and Men.* London: J. Murray, 1820.

Stadler, Rob. *The Scientific Approach to Evolution: What They Didn't Teach You in Biology.* North Charleston: CreateSpace Independent Publishing Platform, 2016.

Statham, Dominic. "The Truth about the Galileo Affair." *Creation.com* (November 8, 2018). https://creation.com/galileo-church.

Steinhardt, Paul. "The Inflation Debate: Is the Theory at the Heart of Modern Cosmology Deeply Flawed?" *Scientific American* 304 (April 2011) 18–25. http://dx.doi.org/10.1038/scientificamerican0411-36.

Strogatz, Steven. *The Calculus of Friendship: What a Teacher and a Student Learned about Life While Corresponding about Math.* Princeton: Princeton University, 2009.

Strobel, Lee. *The Case for The Resurrection: A First-Century Reporter Investigates the Story of the Cross.* Grand Rapids: Zondervan, 2010.

Suetonius. *The Lives of the Twelve Caesars.* Public Domain, 1907. https://penelope.uchicago.edu/Thayer/E/Roman/Texts/Suetonius/12Caesars/home.html.

Tacitus. *The Annals.* Translated by Alfred John Church and William Jackson Brodribb. http://classics.mit.edu/Tacitus/annals.html.

Tan, Change Laura, and Rob Stadler. *The Stairway To Life: An Origin-Of-Life Reality Check.* Evorevo Books, 2020.

Ting, H. A. Christopher. "Quantum Mechanics for Mechanical Engineers." Graduation Thesis, University of Tokyo, 1986. http://cting.x10host.com/CT/Bachelor_Thesis.pdf.

———. "Three-Dimensional Image Analysis of Single-Headed Heavy Meromyosin Bound to Actin-Tropomyosin Filament." Master's Degree Thesis, University of Tokyo, 1988. http://cting.x10host.com/CT/Master_Thesis.pdf.

———. "Magnocellular Pathway for Rotation Invariant Neocognitron." *International Journal of Neural Systems* 4 (1993) 43–54. https://doi.org/10.1142/S0129065793000067.

———. "Topological Study of Field Theories: Braid Group and the Fractional Quantum Hall Effect." PhD diss., National University of Singapore, 1993. http://cting.x10host.com/CT/PhD_Thesis.pdf.

———. *Algorithmic Finance: A Companion To Data Science.* Singapore: World Scientific, 2022.

Ting, Christopher, and C. H. Lai. "Spinning Braid-Group Representation and the Fractional Quantum Hall Effect." *Nuclear Physics B* 396 (1993) 429–464. https://doi.org/10.1016/0550-3213(93)90659-D.

Ting, Christopher, and Keng-Chee Chuang. "An Adaptive Algorithm for Neocognitron to Recognize Analog Images." *Neural Networks* 6 (1993) 285–299. https://doi.org/10.1016/0893-6080(93)90022-O.

Tour, James. "The Mystery of the Origin of Life." *Discovering Science*, YouTube channel (2019). https://www.youtube.com/watch?v=zU7Lww-sBPg.

Turek, Frank. "Science Doesn't Say Anything—Scientists Do." *Cross Examed.Org* (November 25, 2009). https://crossexamined.org/science-doesnt-say-anything-scientists-do/.

———. *Stealing from God: Why Atheists Need God to Make Their Case.* Carol Stream: NavPress, 2014.

Uhlenbeck, George Eugene, and Samuel Abraham Goudsmit. "Spinning Electrons and the Structure of Spectra." *Nature* 117 (1926) 264–265. https://www.nature.com/articles/117264a0.pdf.

Ushijima, Bifue. "Understanding the Origin of the Universe: Explaining What Happened before the Big Bang with Inflation Theory." Translated by Tony Atkinson. *University of Tokyo* (May 14, 2015). https://www.u-tokyo.ac.jp/focus/en/features/f_00066.html.

Vilenkin, Alexander. "Creation of Universes from Nothing." *Physics Letter B* 117 (1982) 25–28.

Vyse, Stuart. *Superstition: A Very Short Introduction.* Oxford: Oxford University Press, 2020.

Wade, Nicholas. "Scanning an Ancient Biblical Text That Humans Fear to Open." *New York Times* (January 5, 2018). https://www.nytimes.com/2018/01/05/science/biblical-codes-morgan-library.html.

Wallace, J. Warner. *Cold-Case Christianity: A Homicide Detective Investigates the Claims of the Gospels.* Colorado Springs: David C Cook, 2013.

———. *Person of Interest: Why Jesus Still Matters in a World that Rejects the Bible.* Grand Rapids: Zondervan, 2021.

Weinberg, Steven. "The Cosmological Constant Problem." *Reviews of Modern Physics* 61 (1989) 1–23.

———. *The First Three Minutes: A Modern View of the Origin of the Universe.* Updated ed. New York: Basic Books, 1993.

Wilkins, John S. *Defining Species: A Sourcebook from Antiquity to Today.* New York: Peter Lang, 2009.

Wilson, A. N. *Charles Darwin: Victorian Mythmaker.* London: Harper, 2017.

Witze, Alexandra. "Four Revelations from the Webb Telescope about Distant Galaxies." *Nature* 608 (2022) 18–19. https://doi.org/10.1038/d41586-022-02056-5.

Wolf, Emil. "Non-Cosmological Redshifts of Spectral Lines." *Nature* 326 (March 1987) 363–365.

Worthington, J. "*Creatio ex Nihilo* and Romans 4.17 in Context." *New Testament Studies* 62 (2016) 49–59.

Xenophanes. "Fragments of Xenophanes." In *The First Philosophers of Greece*, translated by Arthur Fairbanks, 65–77. London: K. Paul, Trench, Trübner & Co., 1898.

Yukawa, Hideki. "On the Interaction of Elementary Particles. I." *Proceedings of the Physico-Mathematical Society of Japan* 17 (1935) 48–57.

Zhang, Yanxia, and Yongheng Zhao. "Astronomy in the Big Data Era." *Data Science Journal* 14:11 (2015) 1–9. https://datascience.codata.org/article/10.5334/dsj-2015-011/.

Zimmer, Andreas D., et al. "Reannotation and Extended Community Resources for the Genome of the Non-Seed Plant *Physcomitrella Patens* Provide Insights into the Evolution of Plant Gene Structures and Functions." *BMC Genomics* 14 (2013) 498. https://doi.org/10.1186/1471-2164-14-498.

Index

www.ingramcontent.com/pod-product-compliance
Lightning Source LLC
LaVergne TN
LVHW051050220926
848485LV00001BA/64

* 9 7 8 1 6 6 6 7 4 8 2 5 3 *